T0319088

Cooperation for competition

Cooperation for competition

Linking Ethiopian farmers to markets

Gian Nicola Francesconi

International chains and network series – Volume 5

Wageningen Academic Publishers

ISBN 978-90-8686-092-0
ISSN 1874-7663

First published, 2009

Wageningen Academic Publishers
The Netherlands, 2009

Acknowledgements

Usually foreign PhD students at the Development Economics Group of Wageningen University study the problems of their own countries. However, this book is not about Italy, but rather about its former 'ghost' colony Ethiopia, invaded for prestige, rapidly converted to spaghetti and macchiato, and abandoned immediately after, left almost completely to solve its own problems. This thesis is the result of 30 months residency in Addis Ababa, and many days spent in remote rural areas, interviewing households with the help of Ato Seifu, Ato Mekonnen, Federico Frosini and Sabine Hiller. This thesis is also the result of many hours spent discussing (inter)national policies, research approaches and methodologies, and much more, with my colleagues at IFPRI-Addis Ababa: Tanguy Bernard, Alemayou Seyoum Tafesse, David Spielman, Eleni Gabre-Madhin, Kassu Wamisho, Zeleka Paulos, Yodit Beyene, Ben Ashenafi, and many more. The period spent in Ethiopia under the dynamic and intelligent supervision of Eleni, and the constant support of outstanding and committed researchers as Tanguy, Alemayou and David has been without doubt the most important experience of my life so far. This was made possible thanks to the financial and institutional support of IFPRI, one of the few intenrational organisations that manage to combine sound scientific research activities with high level policy dialogue and support. Research activities in Ethiopia were co-sponsored by SNV (Netherlands Development Organisation) under the supervision of Dr. Greiling and Mr. Koch, who gave me the opportunity to discuss my preliminary findings in a series of workshops with public and private stakeholders.

The preparation of this thesis involved also a bit less than two years of residency at Wageningen University, under the supervision of Ruerd Ruben, Nico Heerink and Arie Kuyvenhoven. I would like to thank Ruerd for his inspiring ideas, his constant enthusiasm, his lack of political correctness, and his caring supervision. I wish Italy had more professors like you Ruerd! Nico, thanks for guiding me out of the methodological tunnel and for showing me the way towards academic recognition. Arie, I believe your international experience and networking capacity has contributed to place this research in the midst of high level policy debates. Among my numerous colleagues in Wageningen, I would like to mention also the contributions of Marijke D'Haese, Jos Bijman, Gonne Beekman, Ingrid Lefeber, Henny Hendrix, Lonneke Nillesen, Michele Nori, Ricardo Fort, and Marrit van der Berg. In Wageningen I had the chance to interact and learn also from many other students, professors and researchers, who I believe form an academic institute of which I am proud to have been a part. I hope that Wageningen University will remain a place for international students and intercultural discussion, which may indeed bring intangible benefits to the Dutch academia and society, but perhaps these are exactly the type of inputs that are needed in such a productive and efficient country. I also hope that Wageningen University will keep developing cutting-edge interdisciplinary research aiming at the formation of well rounded decision makers, rather than just at the production of highly specialised publications.

A special mention goes to my family (Silvia, Marino, Martina, Giulio, Minerva, Rosolina, Osvaldo and Zoe), who has provided me with unconditional support throughout my lenghty academic career. To my life-time friends Grillo (whose talent and creativity produced the logo of this thesis and the one before) and Christian (whose vision and leadership convinced me to pursue an international career). And to my 'new' friends Tassos and Jacoba, who filled my days in Wageningen with unlimited empathy, deep philosophical thoughts, and radical environmental ideology. Fleur, your contribution to this thesis and to my personal development make me believe that together we could really change the world...and in case we will not make it, I hope that we could just keep enjoying such a wonderful dream.

Cooperation for competition

Table of contents

List of tables

List of figures

'If one only had something to eat,
just a little, on such a clear day!'
Knut Hamsun, 'Hunger', 1890

Chapter 1. Introduction

1.1 Global scenario

'Agriculture is a vital development tool for achieving the Millennium Development Goal that calls for halving by 2015 the share of people suffering from poverty and hunger...Three out of every four poor people in developing countries live in rural areas, and most of them depend directly or indirectly on agriculture for their livelihoods... In much of Sub-Saharan Africa, agriculture is a strong option for spurring growth, overcoming poverty, and enhancing food security...Today, rapidly expanding domestic and global markets; institutional innovation in markets and collective action...offer exciting opportunities to use agriculture to promote development...'

(Robert B. Zoellick, President of the World Bank Group,
World Bank 2007: foreword).

With the declining role of the state, rural development efforts have been gradually shifting from direct aid towards the promotion of employment and entrepreneurship. Assisting rural smallholders to participate in the market is increasingly seen as a sustainable approach to the longstanding problem of global malnutrition and poverty (Fafchamps, 2005; Reardon and Barret, 2000; Cook and Chaddad, 2000; Von Braun, 1995). Nonetheless, the promotion of competitive rural business is a big challenge, especially in Africa.

Several studies (Reardon *et al.*, 2006; Weatherspoon and Reardon, 2003; Delgado *et al.*, 1999) document that in developing countries, food demand for high-value primary products (dairy, meat, horticulture, etc.) is growing rapidly, driven by rising incomes, urbanisation, trade liberalisation, as well as by industrial development and retail concentration into supermarkets. These trends have fostered increasing integration of farms and firms into supply chains in an effort to link rural perishable supply to (inter)national urban demand (World Bank, 2007: 118). Participation in integrated supply chains has the potential to open up new market opportunities for rural smallholders. However, in supply chains, market power tends to concentrate into industrial-retail oligopolies/oligopsonies, posing special challenges to smallholder competitiveness, both in terms of quality and price specifications (World Bank, 2007: 118; Eagleton, 2006; Reardon *et al.*, 2006; Weatherspoon and Reardon, 2003; Kaplinsky and Morris, 2001).

Still, storable agricultural commodities, such as staple cereals (wheat, maize, etc.) and traditional export commodities (coffee, cocoa, tea, cotton, etc.) remain a mainstay for a major share of the rural population (World Bank, 2007: 118). For this reason, and due to increasing world prices, many developing countries have witnessed increasing efforts to reorganise and integrate traditional spot markets into more competitive and trustworthy networks for agri-commodity exchange (COMESA[1]; World Bank, 2007: 120; UNCTAD[2]). However, agri-commodity exchange by smallholders is often hampered because of poor rural infrastructure (roads, telecommunication, etc.) and highly fragmented markets, resulting in high transaction costs and price volatility.

Throughout history rural smallholders have formed various forms of associations (or cooperatives) to confront access-barriers to the market (Staatz, 1987; Sexton and Iskow, 1988; World Bank, 2007: 154). In industrial countries, producer organisations have been fundamental to the success of family farms. In the United States, dairy cooperatives control about 80 percent of dairy production, and most of the specialty crop producers in California are organised in cooperatives. In France, nine of ten producers belong to at least one cooperative, with market shares of 60 percent for inputs, 57 percent for output, and 35 percent for processing (World Bank, 2007: 154).

In the 1960s, many developing countries initiated cooperative development programs, often to facilitate the distribution of subsidised credit and inputs. However, as cooperatives were largely government controlled and staffed, they were often considered as an extended arm of the public sector, rather than institutions or firms owned by the farmers. This form of cooperatives was rarely successful. Political patronage and interference generally resulted in poor performance, corruption and conflicts, which contributed to discredit the movement. This situation began to change in the 1980s as policy reforms embarked on the gradual disengagement of the state from many productive functions and services. The expectation was that removing state control cooperatives could improve production efficiency and quality. Too often, that did not happen. In some countries the state's withdrawal was tentative at best, but not convincing. In others, collective entrepreneurship emerged only slowly and partially (World Bank, 2007: 154).

Still, it is estimated that 250 million farmers in developing countries participate in agricultural cooperatives. Among the better known producer organisations are the Indian Dairy Cooperatives Network and the National Federation of Coffee Growers of Colombia. In 2005 the Indian Dairy Cooperatives, with 12.3 million members, accounted for 22 percent of the milk produced in India. Sixty percent of members in this cooperative are either landless, smallholders, or women (women make up 25 percent of the membership). Created in 1927, the National Federation of Coffee Growers of Colombia has 310,000 members, most of them smallholders (less than 2 hectares). This Federation of cooperatives uses its revenues to contribute to the National Coffee Fund, which finances research and extension and invests

---

[1] COMESA Agricultural Programmes: Executive summary. http://www.comesa.int/agri/Folder.2005-09-12.2953/Multi-language_content.2005-09-12.3013/view

[2] Developing a Pan-African Commodity Exchange. www.ifsc.co.bw/docs/speech_UNCTAD.pdf.

in services (education and health) and basic infrastructure (rural roads, electrification) for coffee-growing communities (World Bank, 2007: 155).

Many donors and governments consider agricultural cooperatives to be a fundamental pillar of development policies, as well as a core institution in the process of governance decentralisation and business development (World Bank, 2007: 155). Many firms consider cooperatives as business partners. For these reasons, a return to agriculture (see World Bank, 2007: 155) coincided with a return to agricultural cooperatives. However, many lessons remain to be learned, and literature reports varying levels of success for agricultural cooperatives in developing countries (see Neven *et al.*, 2005; Chirwa *et al.*, 2005; Sharma and Gulati, 2003; Damiani, 2000; Uphoff, 1993; Attwood and Baviskar, 1987; Tendler, 1983).

1.2 Ethiopian setting

Agriculture is the backbone of the Ethiopian economy, contributing to 48 percent of the gross domestic product (World Bank, 2007: 340). With 85 percent of the population (75 million) living in rural areas under subsistence or semi-subsistence regimes (Alemu *et al.*, 2006; CSA, 2000) and favourable agro-ecological conditions (Ahmed *et al.*, 2003), Ethiopia needs to generate agricultural growth (Gabre-Madhin and Goggin, 2005; Gabre Madhin, 2001).

Although national sources report a recent (since 2003) acceleration in agricultural growth, and especially in crop production per capita (Taffesse *et al.*, 2006), the figures reported by international agencies (FAOSTAT[3]) appear to be less optimistic. In particular, cereal production per capita is showing wide fluctuations, but no clear growth over time (Figure 1.1). Milk production per capita stepped up in 2000, but since then is stagnating (Figure 1.2) and remains clearly below the figures reported from neighbouring Kenya and the whole African continent (Ahmed *et al.*, 2003). Although Ethiopian coffee remains a top quality product in the world (Ethiopia is the birthplace of the bean), coffee production per capita is decreasing steadily (Figure 1.3). Among the agricultural products considered in this study, only sesame seeds show a clear growth in production per capita over time (Figure 1.4).

To revitalise agricultural growth, the Ethiopian government and various international donors approved, in 2006, the proposal of the International Food Policy Research Institute (IFPRI) to establish and launch the first Ethiopian Commodity Exchange (ECX) by 2008. As explained by Gabre-Madhin and Goggin (2005), the ECX is a central marketplace where sellers and buyers meet to transact in an organised fashion, with certain clearly specified and transparent 'rules of the game'. The ECX includes a central trading floor in Addis Ababa connected to warehouses, banks, information and bidding centres in the rest of the country (Gabre-Madhin and Goggin, 2005). The primary goal of the ECX is to promote the commercialisation of major Ethiopian agri-commodities, such as grains, pulses, oil seeds, and coffee. At the same time, due

---

[3] http://faostat.fao.org/

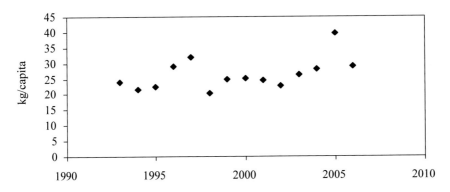

Figure 1.1. Production of cereals, nec, Ethiopia. Source: FAOSTAT.

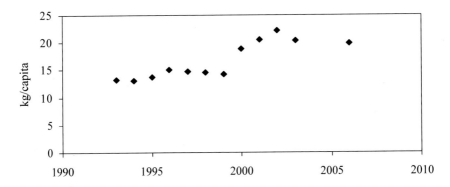

Figure 1.2. Production of raw cow milk, Ethiopia. Source: FAOSTAT.

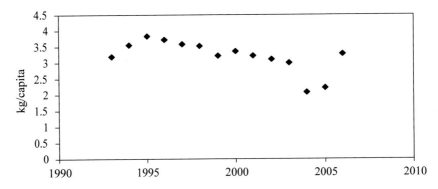

Figure 1.3. Production of green coffee beans, Ethiopia. Source: FAOSTAT.

Cooperation for competition

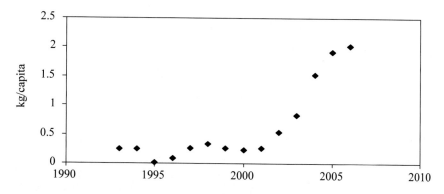

Figure 1.4. Production of sesame seeds, Ethiopia. Source: FAOSTAT.

to rapid urbanisation (World Bank, 2007: 334; CSA, 2004), and market liberalisation reforms (WTO[4]; Ahmed *et al.*, 2003), Ethiopia is witnessing the rapid evolution of the industrial and retail sectors, leading to increasing market integration into supply chains for fresh and perishable food products, and especially for milk (Ahmed *et al.*, 2003).

In Ethiopia, agricultural cooperatives are a pillar of the national strategy named Agricultural Development-Led Industrialisation (ADLI). According to proclamation 85 from 1994 (FDRE), cooperatives in Ethiopia are defined as '...*associations established by individuals on a voluntary basis, to collectively solve economic and social problems and to democratically manage them*'. The role of cooperatives is to mobilise rural entrepreneurship and ensure the participation of smallholders (FDRE, 1998, 1994). Among the better known producer organisations is the Oromia Coffee Farmers Cooperative Union, which exports coffee to the European Union, the United States and Australia.[5]

However, cooperatives are not a novelty in Ethiopia. On the contrary they have a long and controversial history. With the downfall of the Derg regime in 1991, state-owned cooperatives collapsed in many parts of the country symbolizing the liberalisation of farmers from over-centralised governance and corruption (see also Bernard *et al.*, 2008). Nonetheless, cooperatives began to re-emerge after a few years, strongly encouraged and supported by governmental and non-governmental development agencies. As a result the share of *kebeles* with cooperatives went up from 10 percent in 1991, to nearly 35 percent in 2005 (Bernard *et al.*, 2008).[6]

In 2002 the cooperative movement was further reinforced as the Ethiopian government established the Federal Cooperative Commision (FCC) with the ambitious mandate to

---

[4] http://www.wto.org/english/thewto_e/whatis_e/tif_e/org7_e.htm

[5] ILO, http://www.ilo.org/wow/Articles/lang--en/WCMS_081366/index.htm

[6] In Ethiopia the *kebele* is the smallest administrative unit, below the municipality-district level.

provide at least one cooperative per *kebele* to 70 percent of the national *kebeles* by 2010 (FCC, 2006). In 2005, nine percent of the rural households (approximately 6 million individuals) were engaged in some form of collective action. Although agricultural cooperatives have been growing rapidly in Ethiopia, they appear to be still largely dependent on public support (Spielman *et al.*, 2008), while their competitivenes in the market is still raising contrasting evidence (see Bernard *et al.*, 2008; Holloway *et al.*, 2000; Nicholson, 1997).

1.3 Scope of the study

The scope of this study is to improve the understanding of the role played by cooperative organisations in linking Ethiopian smallholder farmers to emerging supply chains and exchange networks. To do so, this study addresses three major research questions.

a. What are the trends and challenges in the Ethiopian food market?

Due to population, urbanisation and economic growth, and increasing efforts towards liberalisation and privatisation, the Ethiopian market is witnessing increasing integration and organisation of stakeholders into supply chains and commodity exchange networks. Although similar trends are reported from several other developing countries, available literature presents little and often out-dated empirical evidence about the socio-economic challenges and opportunities associated with the transformation of the Ethiopian market. To address this gap in research, *Chapter Two* describes the emergence of the first Ethiopian Commodity Exchange (ECX), while *Chapter Four* describes the evolution of national dairy supply chains.

b. What is the impact of collective action for the competitiveness of Ethiopian farmers?

In Ethiopia, like in several other developing countries, agricultural cooperatives are emerging as a means to help smallholders cope with the challenges, and take advantage of the opportunities at local and regional markets. In Ethiopia, like in many other developing countries, agricultural cooperatives operate in the context of rural communities, and therefore they are subject to norms and values of social inclusion and solidarity (World Bank, 2007: 155). Agricultural cooperatives are expected to include and support poor households, and still compete in the marketplace. As such cooperatives have to deal with efficiency versus equity dilemmas.

To what extent Ethiopian cooperatives are able to reconcile business and social purposes remains largely unclear. While Nicholson (1997) and Holloway *et al.* (2000) suggest that Ethiopian dairy cooperatives can serve as a means to increase marketable surplus production by smallholders, Bernard *et al.* (2008) suggest that Ethiopian cereal cooperatives do not include the poorest among rural households, and have a negligible impact on agricultural commercialisation. Although Ethiopian cooperatives rely on cheap family labour their impact on output commercialisation is controversial.

Contrasting evidence about marketing benefits is associated with rare and vague information about the quality of supplies. Too often agricultural cooperatives in Africa are used as instruments to increase (cheap) food provision, no matter if the output is appreciated or not by final consumers. Quality is an essential lubricant of market mechanisms, and minimum hygienic and nutritional standards are undeniable rights of consumers, even when these are undernourished. The impact of collective action on agricultural quality has potential implications for public health and malnutrition, as well as for farmers' competitiveness in the market. Against this background *Chapter Two* assesses the impact of Ethiopian agricultural cooperatives on smallholders' commercialisation processes, and *Chapter Five* measures the impact of collective action on farmers' production and quality.

c. How to promote the competitiveness of Ethiopian agricultural cooperatives?

The impact of agricultural cooperatives on farmers' competitiveness depends on both external and internal factors (Cook and Chambers, 2007; World Bank, 2007: 154). Internal challenges are associated with the heterogeneous interests of an increasingly diverse membership (World Bank, 2007: 155). Conflicts of interest are expected to rise in cooperatives, undermining the stability of the coalition and business performance. *'Whenever the organisational structure of a cooperative is not aligned with the degree of heterogeneity in members' preferences, inefficiencies result in sub-optimal performance and eventually sustainable competitive advantage is forfeited'* (Cook and Chambers, 2007). Cooperative managers are expected to orchestrate an increasingly heterogeneous membership in such a way to meet price, quality and safety specifications of buyers and consumers. Cooperative business requires managerial skills that are often beyond the actual capacity of smallholder farmers in developing countries, justifying external support (World Bank, 2007: 156). Public support should create an enabling environment for cooperative business development, and avoid invasive interventions that can create dependency instead of entrepreneurship. As observed in Ethiopia, cooperatives can be used as instruments to implement policies designed without consulting them, to fulfil the agenda of the donors.

Agribusiness literature presents useful guidelines for cooperatives willing to improve quality management (see Weaver and Kim, 2001), organisational and financial strategies (see Sykuta and Cook, 2001) in the context of industrialised countries with mature markets. Development economics literature instead focuses on the role of governance (see Holloway *et al.,* 2000), typically overlooking intra-cooperative organisation. *Chapters Three and Six* attempt to fill these gaps in research analyzing the relationship between governance, managerial practices, commercialisation and production quality, in Ethiopian agricultural cooperatives.

1.4 Analytical framework

The institutional scenario of the Ethiopian food market can be disentangled into exchange networks for storable commodities, and supply chains for perishable products. The commodity exchange networks described in this study deal mainly with staple cereals, such as teff, wheat, and maize, and traditional export commodities, such as coffee and sesame. The supply chains analysed deal exclusively with milk and dairy products. *Chapters Two* and *Three* address the three research questions discussed above in the context of grain, coffee and sesame markets, while *Chapters Four, Five* and *Six* address the same questions in the context of dairy supply chains. The role of agricultural cooperatives in promoting farmers' competitiveness is also expected to vary across different market scenarios (Table 1.1). This study defines farmers' competitiveness on the basis of productivity, production quality compliance and commercialisation. Output commercialisation is used as an indicator of farmers' engagement in commodity exchange networks. Productivity and production quality are instead used as indicators for farmers' engagement in dairy chains.

The lengthy lag between the decision to plant a crop and the achievement of an output means that market prices at point of sale are unknown at the time decisions are made. The problem is more severe where information is lacking and markets are imperfect (i.e. market affected by high transaction costs), features that are prevalent in Ethiopia. According to Gabre Madhin (2001) and Gabre Madhin and Goggin (2005), Ethiopian grain markets are affected by high price volatility that encourages farmers to produce for home (self) consumption rather than for commercial purposes. Price fluctuations can be particularly severe for perennial tree crops (such as coffee) with a lag of several years between planting and harvest, and less important for daily productions (such as dairy). Unlike perennial tree crops and staple crops (such as grains), productions characterised by faster turn-over (such as dairy) are typically traded under contractual agreements or through vertically integrated supply chains, in which price volatility is less of an issue, but productivity and quality performance are key for competition.

Agricultural cooperatives can be cost-saving and risk-sharing devices for farmers in uncertain agri-commodity markets. Several authors argue that the potential advantages of cooperative farming in generating economies of scale and scope contribute to reduce transaction costs, and to improve bargaining power vis-à-vis the market (Bonin *et al.*, 1993; Munckner, 1988; Dulfer, 1974). However, agri-business literature emphasises the complexities added when multiple individuals, rather than a single investor, engage in commercial activities (Cook and Chambers, 2007; Cook, 1995; Putterman and DiGiorgio, 1985; Vitaliano, 1983; Fama, 1980;

Table 1.1. Analytical framework of the study.

Market scenarios	Commodity exchange (wheat, teff, maize, sesame, coffee)	Supply chains (dairy)
Research question: *a*	Chapter: 2	Chapter: 4
Research question: *b*	Chapter: 2	Chapter: 5
Research question: *c*	Chapter: 3	Chapter: 6
Cooperative role	To promote commercialisation	To promote productivity and quality

Jensen and Meckling, 1976; Olson, 1965). Moreover, cooperatives in developing countries are typically village-level, community-based organisations that face considerable difficulties in combining social equity purposes with commercial activities (World Bank, 2007: 155). Based on similar arguments, Bernard *et al.* (2008) suggest that the formation of cooperatives provides no clear advantages for the commercialisation of grains in Ethiopia.

According to Holloway *et al.* (2000) and Ahmed *et al.* (2003), the production of marketable milk surplus by Ethiopian farmers increases with the adoption of high yielding, cross-bred cows. However, hybrid cows are usually less resistant to diseases than indigenous zebu cattle, and milk production intensification is often associated with a reduction in milk nutrient density (Balasini, 2000; Taneja, 1999). Ethiopian dairy cooperatives can facilitate access to cross-bred cows (Spielman *et al.*, 2008; Holloway *et al.*, 2000), but their role in monitoring and enforcing milk quality standards is often unclear. In Ethiopian dairy cooperatives the milk supplies collected from the farmers are evaluated on the basis of simple field tests or sensorial perceptions, and the use of concentrated feed to support production intensification is rare (Tegegne, 2003).

Because of either productivity or quality shortfalls, many agricultural cooperatives in developing countries cannot meet the specifications of emerging supply chains (World Bank, 2007: 156). While the neoclassical approach (Helmberger and Hoos, 1995; Nourse, 1945) suggests that cooperatives can compete with investor-owned firms, other research building on agency and game theory suggests that traditional cooperative principles undermine optimal allocation of resources and investment policies (Vitaliano, 1983). Major problems of agricultural cooperatives appear to be related to heterogeneous membership occasioning free-riding behaviour and discouraging investments and capital mobilisation due to horizon problems (Putterman and DiGiorgio, 1985). Cooperatives face major challenges in terms of agency coordination, and are notably deficient in providing adequate incentives to prevent free-riding behaviour (Fama, 1980), and in mobilizing equity capital towards production upgrading and intensification (Cook, 1995; Jensen and Meckling, 1976).

On the basis of this analytical framework, the study presents a description of the characteristics and evolution of both commodity exchange networks and integrated value chains in Ethiopia; it evaluates the impact of collective action in promoting farmers' competitiveness, and identifies some policy and managerial options to improve the role of national cooperative organisations in both market scenarios.

1.5 Data and methods

In this study, existing theories on cooperative impact and management are tested against primary data from Ethiopia. This study makes use of five datasets resulting from five consecutive surveys conducted between 2003 and 2006. These surveys included both households and cooperatives as unit of analysis and were conducted using structured questionnaires. In addition to quantitative data, the analysis includes information gathered through semi-structured interviews, and open discussions with farmers, cooperative managers, representatives from the private sector (Tetra Pack, and local entrepreneurs), and officers of the Ethiopian Federal Cooperative Commission (FCC), the Ethiopian Development Research Institute (EDRI), the Line Ministries, and international development agencies (ILRI, IFPRI, IWMI, ECA, World Bank, and various NGOs).[7]

The use of relatively small datasets (100-400 observations) confined to particular regions and collected at one point in time does not allow for the drawing of more general conclusions. Lack of longitudinal time-series data affects the capacity to capture the direction of causality in the events analysed. However, the intrinsic limitations of relatively small and cross-section datasets are partly compensated by the direct involvement of the author in questionnaire design, data collection and interviews. These conditions ensure a thorough understanding of the data available, facilitate the detection and control of measurement errors, and allow for the back-up of empirical results with direct observations and *vice versa*. Although the number of observation per data set is fairly small, this study uses five separate datasets adding up to a total of 1,100 observations, providing a pretty comprehensive overview of economic, managerial and biological aspects of the agro-industrialisation process in the Ethiopian Highlands.

Data and information were processed using several analytical methods. *Chapter Two* (in addressing *Research Question a*) draws from existing evidence and theory on commodity exchange networks (Gabre Madhin and Goggin, 2005) to describe the characteristics and dynamics of the emerging ECX. *Chapter Four* (in addressing *Research Question a*) elaborates on the econometric and qualitative methods proposed by Neven *et al.* (2006), and the market power analysis proposed by Kaplinsky and Morris (2002), to evaluate evolutionary patterns and challenges in Ethiopian dairy supply chains. *Chapter Two and Five* (addressing *Research Question b*) draw from the analytical approach elaborated by Ravallion (2001), and use quantitative methodology based on both propensity score matching and regression analysis (and several variations of each method), as proposed by Godtland *et al.* (2004), to evaluate the differences in commercialisation, productivity and quality between cooperative farmers and control groups of otherwise similar farmers.

Chapter Three (addressing *Research Question c*) elaborates on the cutting-edge life cycle framework from Cook and Chambers (2007), which models the evolution of business performance of US agricultural cooperatives. The life cycle theory suggests that cooperative business is characterised by an initial stage with high turnover, followed by a steady reduction in sales due to increasing market competition. This study provides a detailed analysis of business life cycles in Ethiopian agri-commodity cooperatives, and shed some light on the timing

---

[7] IFPRI is the International Food Policy Research Institute (CGIAR). ILRI is the International Livestock Research Institute (CGIAR). IWMI is the International Water Management Institute (CGIAR). ECA is the Economic Commision for Africa (UN). NGOs include SNV (the Netherlands), Land' o Lakes (USA), LVIA (Italy), and the Bilateral Italian Cooperation of the Ministry of Foreign Affairs (Italy), among others.

and type of interventions needed to promote and sustain collective commercialisation by smallholder farmers. *Chapter Six* (addressing *Research Question c*) draws from the institutional framework elaborated by Weaver and Kim (2001) to identify key cooperative arrangements that can provide Ethiopian farmers with incentives for milk quality upgrading.

1.6 Relevance of the study

Cooperatives are the backbone of the Ethiopian agricultural policy. Ethiopian cooperatives are social institutions that exist for mutual support purposes, as well as firms aiming at profit maximisation. In the literature these two sides of the same coin are commonly addressed in separate fields of inquiry, development economics and agri-business research, which too often resemble 'adventurers following parallel paths' (Reardon and Barret, 2000), or 'boats passing side by side in the night, without noticing each other' (Cook and Chaddad, 2000). Perhaps this is because the paradigm of agroindustrialisation for development is relatively new, or at least its importance was recognised only recently. Or perhaps it is because both development economics and agri-business research are relatively new fields, and energies are still focused on understanding intra-field complexity, thereby augmenting opportunity costs of adventuring beyond paradigmatic borders.

Whatever may be the cause, in the last decade agri-business research has begun to raise global discontent, largely related to the excessive attention paid to profit maximisation strategies leading to unequal distribution of wealth and power. On the other hand, development economics has specialised to serve a diminishing public clientele, as if the private sector had no social responsibility. Consequently, rural smallholders remain largely dependent from a diminishing public support and often excluded from emerging markets. This leads to the exacerbation of the gap between a few urban rich and rural many poor, fuelling socio-political instability, which in turn undermines the agroindustrialisation process. This study aims at raising awareness of the potential of multidisciplinary research, combining agri-business and development approaches, as a guide for public-private partnerships in promoting collective marketing in developing countries.

'A cooperative is not a cooperative
is not a cooperative is not a ...'
Michael Cook, Missouri University, 2007.

Chapter 2. Linking smallholders to commodity exchange: the role of agricultural cooperatives in Ethiopia

Abstract

The government of Ethiopia is actively promoting the involvement of cooperatives in the newly established commodity exchange. Using household survey data collected in 2005, this study evaluates the impact of smallholders' cooperatives on agri-commodity (teff, maize, wheat, sesame, and coffee) comercialisation in rural Ethiopia. To do so we examine the factors explaining the degree of commercialisation of cooperative farmers and individual farmers located in major agri-commodity production sites. To eliminate potential diffusion effects between cooperative farmers and farmers that do not belong to cooperatives, we select the latter from comparable communities with no cooperatives. Findings from Tobit regression and propensity score matching are consistent across the two methods in suggesting that cooperative membership has an insignificant impact on agri-commodity commercialisation. Only cooperatives that engage in collective marketing activities, such as the collection and sale of members' output, appear to have a significant and positive impact on smallholder commercialisation. The study concludes with implications for policy and for further research.

2.1 Introduction

Ethiopia is the largest producer of maize and wheat in Africa, with domestic production more than double the volumes jointly produced by Kenya, Tanzania and Uganda in 2004-05 (Gabre-Madhin and Goggin, 2005). Ethiopia is also Africa's largest coffee producer and the birthplace of the bean. Overall, grains, coffee and other agri-commodities, such as pulses and oil seeds, are central to the Ethiopian economy, engaging almost 10 million smallholder farmers, and related households, in the production process. In Ethiopia, like in many other African countries, agri-commodity commercialisation has the potential to boost the economy and reduce poverty.

Despite the downfall of the Derg regime in 1991, and subsequent policy reforms towards market liberalisation, numerous studies (Dadi *et al.*, 1992; Lirenso, 1993; Dercon, 1995; Negassa and Jayne, 1997; Dessalegn *et al.*, 1998; Gabre-Madhin, 2001) document that agri-commodity flow from rural Ethiopia to (inter)national urban markets remains highly constrained. Why? Policy reforms largely overlooked the marketing problems faced by smallholder farmers whose production constitutes the bulk of the domestic agricultural produce. Like in many other transition economies, vanishing governmental control over prices and outlets left Ethiopian smallholder farmers in a state of increased uncertainty (World Bank, 2007: 138).

Where to find a buyer? What price to expect? What type of quality standards and exchange rules to adopt? Increasing transaction costs came along with the exacerbation of price volatility (Gabre-Madhin and Goggin, 2005; Dercon, 1995). As a result, the benefits linked to market liberalisation were overshadowed by growing economic uncertainty, and subsistence and semi-subsistence farming remained dominant all over rural Ethiopia (Alemu *et al.,* 2006; Alemu and Pender, 2007).

In response to this problem, the Ethiopian government and various international donors approved in 2006 the proposal of the International Food Policy Research Institute (IFPRI) to establish and launch the first Ethiopian Commodity Exchange (ECX) by 2008. The primary goal of the ECX is to promote the commercialisation of major agricultural commodities, such as grains, pulses, oil seeds, and coffee. As explained by Gabre-Madhin and Goggin (2005), a commodity exchange is a central marketplace where sellers and buyers meet to transact in an organised fashion, with certain clearly specified and transparent 'rules of the game'. In its wider sense, a commodity exchange is any organised marketplace where trade is funnelled through a single, well defined mechanism. A commodity exchange is expected to increase trust among buyers and sellers, a necessary first condition for trade to occur. Based on modern information and communication systems, a commodity exchange is also expected to increase the concentration of buyers and sellers over a single trading floor, improving effective market competition. If this is the case, the trade mechanism based on price bidding or auctions will results in what is known as 'price discovery', that is, the emergence of the true market-clearing price for a good at a particular point in time.

In brief, a commodity exchange is an institutional response to the fundamental problem of 'thin markets', defined as markets in which there are few purchases and sales. Although the ECX represents a very innovative solution for Ethiopia, indeed it is nothing new or unexpected. The ECX fits in the plans of the Common Market for Eastern and Southern Africa (COMESA), which has expressed interest in forming a regional commodity exchange (commex), taking up from growing initiatives observed at the national level in several member states (including Kenya, Uganda, Zambia, Zimbabwe, Malawi, South Africa, besides Ethiopia). Overall, the increasing efforts displayed in various parts of Africa to institutionalise agri-commodity exchange follows the successful experiences of the Chinese Commodity Exchange (created in the early 1990s), and the Indian Multi-Commodity Exchange (founded in 2002), in turn inspired by well consolidated examples like the Chicago Board of Trade and the Tokyo Grain Exchange, among others.[8]

---

[8] UNCTAD, 2006. Developing a Pan-African Commodity Exchange. www.ifsc.co.bw/docs/speech_UNCTAD.pdf.

The establishment of a commodity exchange in Ethiopia is expected to create more linkages and tighter integration across rural, urban and export commodity markets, with clear benefits for farms, firms, and consumers. However, easy enthusiasm is largely repressed by widespread concerns related to the outreach potential of the ECX. Ethiopia counts approximately ten million rural households, producing grains, pulses, oil seeds and coffee, mainly for their own subsistence. Lack of capital, remoteness, poorly developed roads and telephone lines are only some of the barriers that keep farm households far away from markets, and therefore from the potential benefits of the ECX. In an era of commodity exchange globalisation, Ethiopian smallholder farmers may remain, once again, at the margin of economic development, consolidating or even exacerbating the gap between rich and poor. To avoid this scenario, the Ethiopian government is strongly promoting the formation of agricultural cooperatives all over the national territory, and their direct involvement in the ECX network. In Ethiopia, agricultural cooperatives are increasingly seen as preferential interfaces between agri-commodity farmers and output markets, and thus as key institutional partners for the upcoming ECX.

During the past decade, donors and governments have regained interest in cooperative mechanisms to overcome barriers to smallholder commercialisation (Collion and Rondot, 1998; World Bank, 2003; Cook and Chambers, 2007). However empirical evidence suggests varying levels of success for cooperatives in developing countries (Chirwa et al., 2005; Neven et al., 2005; Sharma and Gulati, 2003; Damiani, 2000; Uphoff, 1993; Attwood and Baviskar, 1987; Tendler, 1983). With regard to the Ethiopian context, Bernard et al. (2007) report that the formation of smallholder cooperatives provides no clear advantage for the commercialisation of grains. Bernard et al. (2008) find that cooperatives do provide better prices to their farmers, but price incentives are not sufficient for all farmers to ensure greater market participation. In particular, due to the higher prices obtained through the cooperative, poorer members tend to sell less and consume more cereals, whereas richer members, who have larger supply elasticities and smaller income elasticities of cereal consumption, tend to sell more. According to these authors, such heterogeneous responses among cooperative members explain the lack of an aggregate impact of Ethiopian cooperative membership on grain commercialisation.

In this study we argue that beyond such individual capacities and preferences there are particular characteristics of cooperatives, i.e. differences in collective organisational structures, that may also play an important role in determining the commercial behaviour of Ethiopian farmers. Hence, not only heterogeneity among members, but also heterogeneity among cooperative organisations needs to be taken into account. The purpose of this study is to get more insights into the impact of different forms of agricultural cooperatives on agri-commodity commercialisation in the regions targeted by the ECX. In particular this study distinguishes between offensive or marketing cooperatives and multipurpose or defensive cooperatives. This distinction is explained and motivated in section two. In section two, we explore also the characteristics of the ECX and provide a better understanding of motivations and mechanisms underlying the proposed partnership between the ECX and the agricultural cooperatives. In section three we derive from a large dataset, representative of national rural households (Ethiopia Rural Smallholder Survey, IFPRI, 2005), a sub-sample including only major agri-commodity production areas, primary targets of the ECX. Descriptive statistics of the household characteristics and degree of commercialisation for farmers belonging to cooperatives and farmers who are not engaged in cooperatives are also presented. In section four, we present the propensity score matching methodology used in this study to estimate the impact of cooperative membership on smallholders' commercialisation, and the Tobit model that we use to assess the robustness of the results. In section five, we present and discuss the results of our empirical analysis, and contrast them with the available literature. Finally, in section six we summarise our main findings and we discuss the implications for policy and for further research.

2.2 ECX and cooperatives

The ECX deals with six agricultural commodities: teff (the national staple), wheat, maize, coffee, sesame seeds and pea beans. The initial structure of the ECX includes a central trading floor located in Addis Ababa, plus 20 terminal centres and 10 warehouses located in strategic agri-commodity markets (see Figure 2.1). The ECX model comprises also three clearing banks, with offices in all urban areas. The functioning of the ECX is based on modern information and communication technology, and can be essentially distinguished into the following five phases:

a. Sellers place *offers* and buyers place *bids* at a nearby terminal centre or directly at the central trading floor. To do so buyers have to deposit the amount of money bidded in one of the three clearing banks, while sellers have to store the supplies offered in the closest warehouse in exchange of a receipt. At the warehouse level, quality, safety and volume of supplies received will be systematically graded and recorded.

b. Offers and bids are transmitted and stored in a central database, screened in chronological order, and matched only when identical.

c. When offer and bid match, a clearing message is transmitted to the bank where the buyer deposited the funds. The bank simultaneously releases funds to the seller and a receipt to the buyer (to reduce delivery risk).

d. Finally, the price of the completed transaction is transmitted and publicly displayed in 200 *woredas* (Ethiopian municipalities).

In order to ensure that the exchange rules are followed, the model functions with membership-based trading (Gabre-Madhin and Goggin, 2005). Since chaos would quickly result under unlimited membership, the ECX will initially involve 150 members. In addition to an annual fee, the actual seat on the exchange floor has to be bought, and therefore serves like a share that can be bought or sold on the market. This ensures that members have a stake in the performance of the ECX and thus uphold its trust and integrity. To maximise the concentration of buyers and sellers under the ECX system, members are expected to serve as

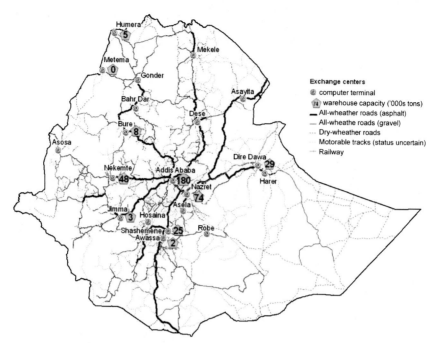

Figure 2.1. ECX structure, 2007. Source IFPRI-ESSP.

brokers. Brokerage is the key mechanism of the ECX, with members trading on behalf of an unlimited number of clients. The function of members is to advise their clients, whether to buy or sell and when market opportunities are likely to occur. Because of their central role, the integrity and capacity of members is at the core of the exchange itself. Therefore, members need to be inspected in their function. Exchange regulations make it mandatory for members to use standard contracts describing the type, origin, quantity and quality of the supply (as certified at the warehouse level), final price of transaction, storage, handling and insurance costs, delivery date, consequence of non-performance and force majeure.

According to the Ethiopian government, the selection of ECX's members should encourage and favour agricultural cooperatives over individual producers and traders. The governmental proposal is motivated by the assumption that the physical presence of cooperative managers on the ECX's trading floor can promote the participation of smallholder farmers in the ECX system, with consequent benefits for poverty alleviation and food security. Still, such a proposal appears questionable. As documented by Bernard *et al.* (2008), agricultural cooperatives have a controversial history in Ethiopia. During the Derg regime (1974-1991), agricultural cooperatives proliferated all over Ethiopia, often imposed by the state to control farmers (a vital source of political support and agricultural products). However, due to growing internal corruption and conflicts, most Ethiopian cooperatives dissolved together with communist ideals. Nonetheless, in 1994, the recently settled government expressed renewed interest in cooperatives as a means to promote the participation of smallholders in input and output markets (FDRE, 1994 and FDRE, 1998). Consequently, since 1994, cooperatives have been actively promoted all over the country, on the basis of three sets of public incentives: (a) start-up incentives, covering all unredeemable establishment costs related to search and screening of potential members, set-up of cooperative boards and commitees, training on cooperative principles and law, legalisation process, early-phase monitoring and technical support; (b) production incentives, in the form of subsidised inputs and preferential access to land (to be rented from the government); and (c) marketing incentives, in the form of facilitated group lending schemes, managerial training and support (in cash or kind), and preferential access to NGO's support, to increase farm-output commercialisation. According to Bernard *et al.* (2008) the share of *kebeles* with cooperatives went up from 10 percent in 1991, to nearly 35 percent in 2006.[9] In 2002 cooperative governance was even reinforced by the establishment of the Federal Cooperative Commision (FCC), a governmental body with the ambitious mandate to establish one cooperative per *kebele* to 70 percent of the national *kebeles* by 2010 (FCC, 2006).

According to Ethiopian proclamations 85 from 1994 cooperatives are defined as '*associations established by individuals on a voluntary basis, to collectively solve economic and social problems and to democratically manage them*'. In order to register with the FCC and achieve legal recognition a cooperative should have at least ten members, who must show no major irregularities in their financial records. Cooperative law states also that any individual has the right to join in, as long as he/she can afford to pay eventual entrance fees, and to purchase at least one share of the equity capital accumulated up to that moment by the group. While the distribution of property rights among cooperative members has been completely deregulated by a subsequent proclamation (number 147 from 1998), decision making processes in cooperatives remain tied by law to the principle of one member one vote. As a result, most cooperatives define fees and shares on the basis of regular internal evaluations and members' approval. Unlike fees, shares are usually redeemable but cannot be traded, not even among members. Furthermore, cooperatives that collect and sell the supplies of member-farmers often retain a fixed percentage of sales' revenues as a form of patronage to build up additional equity capital and cover running costs.

In most cases, Ethiopian cooperatives serve smallholders through providing access to subsidised agricultural inputs (such as land, fertilizer, artificial insemination, improved live animals and improved seeds), farming services and donations from the state and from NGOs (see Spielman *et al.,* 2008). To a lesser extent, agricultural cooperatives provide also basic services for output marketing, such as collection and sale of members' supplies (additional services like storage, transportation and manufacturing, are extremely rare). Following the classification proposed by Sykuta and Cook (2001), Ethiopian cooperatives can therefore be distinguished into

---

[9] In Ethiopia a *kebele*, the smallest administrative units, below the municipality-district level.

marketing (or offensive) and non-marketing (or multipurpose or defensive) organisations. Bernard *et al.* (2008) suggest that marketing cooperatives are a minority in Ethiopia.

2.3 Sample and data

The data used in this study were collected in 2005 during a household survey jointly carried out by the International Food Policy Research Institute (IFPRI), the Ethiopian Development Research Institute (EDRI) and the Central Statistical Agency (CSA). The survey focused on smallholders' commercialisation and covered all rural parts of the country, except the Gambela region, and the non-sedentary population in the Afar (three zones) and Somali (six zones) regions. The sampling procedure adopted for the survey was based on the sampling scheme of the Annual Agricultural Survey carried out by the CSA in 2004-05. Out of the 2014 enumeration areas (EAs) covered by the CSA, 293 EAs were randomly selected and surveyed: 95 from Oromiya, 67 from Amhara, 32 from Tigray, 65 from SNNP (Southern Nations, Nationalities and People), 5 from Afar, 11 from Somali, 12 from Benishangul, two from Harari, two from Diredawa, and two from Addis Ababa regions.[10] To maintain representativeness, the number of EAs per region in the survey is proportional to the number of EAs sampled by the CSA in each region.

The questionnaire included seven modules: demographics, crop and livestock production, commercial behaviour, assets, public infrastructures and social services, social capital, and shocks and coping strategies. Due to the large spatial coverage and the household sample size of more than 7,000 units, the commercialisation survey was carried out by 100 enumerators and nine survey experts (or supervisors). The survey was launched in mid May 2005, and was completed within the second week of July 2005.

From the original sample we created a sub-sample to be used in this specific study that comprises only farm households located in *woredas* (i.e. municipalities) hosting ECX centres (Figure 2.1). Such a sub-sample allows us to focus on markets and farms primarily targeted by the ECX. Since three of the target *woredas* (Robe, Harar and Asella) were not included in the original sample, the sub-sample used for our analysis comprises a total of 17 (instead of 20) *woredas*. In each *woreda*, 24-25 farm households were surveyed, giving a total sample size of 417 households (see Table 2.1).

In our sub-sample, 88 percent of the farm households (368 farms) grow at least one of the six commodities of interest to the ECX (i.e. maize, wheat, teff, sesame, coffee and pea beans). To avoid problems related to sample selection bias the following analysis excludes farm households that do not produce any of the ECX-commodities. It is also important to note that none of the 368 households considered happened to grow pea beans during the period investigated by the survey, and therefore this specific ECX-commodity is excluded from the analysis.

---

[10] An average EA includes between 150 to 200 households.

Table 2.1. Sample characteristics, ECX sites, 2005.

ECX centres	Region	Zone	Name & code of *woreda* selected	Sample size
Mekele	Tigray	Southern Tigray	Enderta 10402	24
Humera	Tigray	Western Tigray	Kafta 10501	25
Asayta	Afar	Zone I	Asayta 20103	24
Gonder	Amhara	North Gonder	Gonder Zuria 30111	25
Metema	Amhara	North Gonder	Metema 30114	24
Dessie	Amhara	South Wollo	Dessie Zuria 30410	24
Bahir Dar	Amhara	West Gojam	Bahir Dar Zuria 30703	25
Bure	Amhara	West Gojam	Bure Wemberma 30717	25
Nekempte	Oromia	East Wellega	Guto Wayu 40215	24
Jimma	Oromia	Jimma	Kersa 40405	25
Nazareth	Oromia	East Shewa	Adama 40703	25
Shashemene	Oromia	East Shewa	Shashemene 40712	24
Asosa	Benishangul Gumuz	Asosa	Asosa 60303	25
Hosaina	SNNP	Hadiya	Limu 70202	24
Awasa	SNNP	Sidama	Awasa 70402	24
Addis Ababa	Addis Ababa	AA Zone 6	Wereda I 140317	25
Dire Dawa	Dire Dawa	Dire Dawa	Dire Dawa 150102	25
				417

Out of the 368 farm household analysed, 21 percent are enrolled in at least one cooperative. This figure is surprisingly high compared to the 9 percent reported by Bernard *et al.* (2008) for the whole country, suggesting that the incidence of cooperatives is higher in the major agri-commodity production sites in Ethiopia. Cooperative members are not homogeneously distributed across these sites. Our data show that they are concentrated in eight *woredas*. On the other hand, individual farmers, i.e. farmers who are not a member of a cooperative, are found in each of the 17 *woredas* included in the sample. Approximately half of the cooperative farmers (11 percent) indicate that their cooperatives provide marketing services, i.e collection and sale of farm output.

Table 2.2 compares the characteristics and the degree of commercialisation of individual farm households and cooperative farm households, distinguishing further between overall cooperative farms and farms engaged in marketing cooperatives. It shows that farm households engaged in cooperatives differ significantly from individual farm households in several aspects. In particular, cooperative households are larger (approximately one extra household member); have a higher dependency ratio (the share of children below 14 in the household); have a

Table 2.2. Characteristics and commercialisation of farm households, ECX sites, 2005.

[368 obs.]	Individual farmers [290 obs.]	All cooperative farmers [78 obs.]	Market-coop farmers [42 obs.]
Number of household members	4.87 (2.23)	5.86 (2.23)**	5.71 (2.60)**
Dependency ratio (children/adults)	1.09 (1.09)	1.36 (1.01)**	1.31 (1.12)
Age of household head (years)	43.88 (15.31)	43.36 (13.04)	44.09 (12.26)
Dummy for male household head	0.77 (0.42)	0.91 (0.29)**	0.83 (0.38)
Education of household head (years)	3.02 (6.33)	5.90 (8.44)**	4.98 (7.52)*
Distance from nearest market (min on foot)	75.18 (30.41)	74.52 (37.41)	66.42 (42.83)*
Fixed arable land (hectares)[1]	1.39 (1.31)	2.93 (2.85)**	3.58 (3.64)**
Degree of autarky/commercialisation[2]	0.28 (0.38)	0.42 (0.37)**	0.56 (0.36)**

Standard deviations in parenthesis ().

* denotes significant difference at 10% level, while **denotes significant difference at 5% level, between the mean of individual farmers and the mean of these two categories.

[1] Land in Ethiopia is the property of the state and cannot be owned by farmers. It is allocated to farmers for an undetermined period. Although land cannot be sold, it can be rented out and eventually passed on to heirs. For a detailed description of land tenure system, see Gebreselassie (2006). Land size in this paper refers to the size of the allocated land.

[2] The commercialisation index, c, is computed as the ratio of the value of ECX-commodities sold, vs, to the total value of ECX-commodities produced, vy, by a farm:

$$c = \frac{vs}{vy} \text{ with } vy = \sum_{n=1}^{N} y_n p_n^* \text{ and } vs = \sum_{n=1}^{N} s_n p_n^*$$

where ECX-commodities ($n = 1,2...N$) include teff, wheat, maize, sesame and coffee, y indicates the volume produced, s the volume sold, and p^* the average sample price for teff (2.17 Birr), wheat (1.63 Birr), maize (1.15 Birr), sesame (5.08 Birr) and coffee (11.57 Birr), respectively.

higher incidence of male heads, have more educated heads (approximately three extra years of schooling); and have larger land sizes. Similar differences are observed between individual farms and farms engaged in marketing cooperatives. However, unlike average cooperative farmers, members of marketing cooperatives do not differ significantly from individual farmers in gender of household heads and dependency ratio. Instead they are significantly closer to markets.

Following Von Braun (1995), Strasberg et al. (1999) and Alemu et al. (2006), we measure farm commercialisation as the ratio between the value of ECX-commodities sold and the total value of ECX-commodities produced.[6] Hence, a value of zero indicates a farm household where teff, wheat, maize, sesame or coffee are produced exclusively for home consumption. The closer the index is to one, the higher is the commercialisation of these specific agri-commodities. In other words, this index quantifies the degree of market orientation or access (as opposed to the degree of autarky) of a farm household. Using this index, members of cooperatives, and in particular the members of marketing cooperatives, appear significantly more commercial than individual farmers. The average share of the ECX commodities sold by cooperative farmers is about 1.5 times the average shae sold by individual farmers (42 versus 28 percent). For farm households involved in marketing cooperatives, this share is even twice as large (56 versus 28 percent) on average.

2.4 Methodology

The analytical method used in this study draws from the work of Ravallion (2001), Godtland *et al.* (2004) and Bernard *et al.* (2008). According to these authors, a way to obtain robust impact assessments is to compute the Average Treatment effect on the Treated (ATT), which in this case refers to the average effect of cooperative membership on the degree of commercialisation of cooperative members. The empirical problem we face in this case is the typical absence of data concerning the counter-factual: what would cooperative farmers have done if they had not joined the cooperative? Our challenge is to identify a suitable comparison group of non-participants whose outcomes – on average – provide an unbiased estimate of the outcomes that cooperative members would have had in the absence of the cooperative. Given the non-random selection of cooperatives' location (the establishment of cooperatives depends largely on government interventions), and farmers self-selection into cooperatives (membership is a voluntary decision depending on farm resources, as well as farmer preference), a simple comparison of outcomes between participants and non-participants (i.e. naïve comparison, such as those presented in Table 2.2) may yield biased estimates of cooperative membership impact.

There are three potential sources of bias in naïve comparisons. First, coop-members are likely to differ from individual farmers in the distribution of observable characteristics (such as agro-ecological conditions, public infrastructure and services, market institutions and demands, households characteristics, farm assets and practices, etc.) leading to a bias related to 'selection on observables'. Such a bias is likely to arise because these observable differences can also be expected to have a direct effect on commercialisation in the absence of the cooperative. A second source of bias in cooperative impact can arise in case of diffusion or spill-over effects between cooperatives and the surrounding environment. For instance, a cooperative is likely to attract extension and input services. In many cases the commercial benefits from these services can trickle down to neighbouring farmers that are not members of the cooperative, leading to an underestimation of cooperative membership impact. A third source of bias is that cooperative participants may differ from non-participants in unobservable characteristics (e.g. personal ability, motivations and preference), which may also affect commercialisation, resulting in 'selection on unobservables' or 'self-selection'.

We address these potential sources of bias in the following ways. First, we exclude from the sample all individual farmers located in *woredas* with at least one cooperative. This procedure reduces further the size of the sample, but eliminates any potential sources of diffusion bias. Second, in the absence of a suitable instrument, we are unable to explicitly control for potential bias related to selection on unobservables. However, the strong incentives provided by the government to promote farmers' participation in cooperatives (see section two), give us sufficient reasons to believe that selection on unobservables might also be negligible, especially after the exclusion of individual farmers located in *woredas* hosting cooperatives. In other words, we believe that governmental incentives are sufficient to convince farmers to join, given that cooperatives are available and accessible, and given observable farm household characteristics.

Third, we use the farm household variables presented in Table 2.2 to control for selection on observables. In the absence of reliable data at the community level, we cannot control for location-specific effects associated with market, agro-ecological and infrastructural conditions on the decision to join a cooperative. However, since the *woredas* included in the sample are all considered to be major production and marketing sites in Ethiopia, as well as major sources of agricultural commodities for the ECX system, we assume market, agro-ecological and infrastructural differences across sample sites to be negligible. Hence, we control for potential bias caused by selection on observables using two separate techniques: propensity scores matching (PSM) and Tobit regression analysis. The PSM technique involves the estimation of the propensity of farmers to engage in cooperatives on the basis of farm household characteristics (using Probit models), and subsequently the matching of individual and cooperative farmers on the basis of propensity scores and the estimation of ATT. The Tobit model is used to regress farmers' commercialisation directly on cooperative membership and farm household characteristics. PSM and Tobit techniques allow to control for selection on observables and provide comparable estimations of cooperative membership impact.

In both analyses, endogeneity (i.e. simultaneity) problems are avoided by using explanatory variables that include household and fixed farm characteristics (such as fixed land asset and distance from the market). Moreover, farm-household characteristics are intentionally over-parametrised using quadratic terms in order to take into account possible nonlinearities in the impact of these variables, and to improve the predictions of both analytical models, (see Godtland *et al.*, 2004). A right and left censored Tobit estimator is used as farmers' commercialisation varies between zero and one. The Tobit analyses were tested for the presence of heteroskedasticity (using Breusch-Pagan/Cook-Weisberg test), which appears not significant, and improved through the exclusion of a few influential observations. Statistical robustness of the PSM analysis is instead promoted by matching farmers using two separate techniques (Kernel and Nearest Neighbour), and by comparing the results obtained.[11] To ensure maximum comparability of the treatment and control groups, the sample used for PSM is restricted to the common support region, defined as the values of propensity scores where both treatment and control observations can be found.

2.5 Results

The presentation of our findings starts from the Probit model in Table 2.3 used to compute propensity scores. From this regression emerges that land holding size is the only significant factor in explaining cooperative membership. In particular we notice that the relation between land holding and membership is quadratic and concave. In other words, the probability of being a member of a cooperative increases as the size of allocated land increases, up to a given threshold (approximately 10 hectares) after which the relation becomes negative. As the average land size in our sample is less than two hectares, this finding indicates that larger farmers are more likely to be a member of a cooperative (as also suggested by Bernard *et al.*, 2008). Although small farmers need cooperatives more than large ones to overcome high transaction costs, large farmers appear to have easier access to membership. It is also important to note that the insignificance of all other variables in explaining cooperative membership should be interpreted as a positive result, suggesting that the only observable in which treated and control farmers differ is land size. This confirms the validity of our control group.

The second part of the results (Table 2.4) reveals the average effect of the treatment on the treated (ATT), i.e. the difference in the level of commercialisation between cooperative and individual farmers, after PSM. In particular, Table 2.4 shows that the level of commercialisation of cooperative members does not differ significantly from the level estimated for individual farmers. This finding is in line with the evidence presented by Bernard *et al.* (2008), also suggesting that Ethiopian cooperatives have an insignificant impact on rural commercialisation. However, our analysis suggests also that cooperative members engaged in marketing cooperatives have a significantly higher degree of commercialisation (14-21 percentage points higher) than individual farmers. In other words, our analysis points out that the insignificant impact reported for agricultural cooperatives can be explained from the existence of a sub-group of cooperatives (almost 50 percent of the cooperatives sampled) that does not provide services for output marketing to their members. The commercial impact of the latter sub-group overshadows the positive impact of marketing cooperatives. In brief, the establishment of agricultural cooperatives is not sufficient to link smallholder farmers to agri-commodity

---

[11] Several matching techniques can be used to match treatment and control households. Here we focus on two widely used methods, the non-parametric Kernel regression matching proposed by Heckman (1998), and five-nearest -neighbours matching. In the first case, each treated household is matched with the entire sample of controls. However, each control observation enters the estimate with a weight inversely proportional to its distance to the treatment observation, based on the propensity-score distribution. For the second method each treatment observation is matched with an average value of its five nearest control neighbours, again based on the propensity score distribution.

Cooperation for competition

Table 2.3. Probability of cooperative membership (Probit), ECX sites, 2005.

	All coops	Marketing coops
Fixed arable land (hectares)	0.50 (0.10)**	0.50 (0.12)**
{Fixed arable land}^{2}	-0.02 (0.01)**	-0.02 (0.01)**
Household size (no. of members)	0.01 (0.05)	-0.01 (0.06)
{Household size}^{2}	-1.13 (0.78)	-1.13 (0.79)
Dependency ratio (children/adults)	-0.05 (0.11)	-0.04 (0.12)
Education of household head (years)	0.08 (0.06)	0.12 (0.07)
{Education of household head}^{2}	-0.00 (0.00)	-0.00 (0.00)
Age of household head (years)	-0.01 (0.04)	0.03 (0.05)
{Age of household head}^{2}	-0.00 (0.00)	-0.00 (0.00)
Dummy for male household head	0.35 (0.33)	-0.04 (0.35)
Distance to nearest market (minutes/foot)	-0.00 (0.01)	-0.01 (0.01)
{Distance to nearest market}^{2}	0.00 (0.00)	0.00 (0.00)
No. of observations	279	243
Pseudo R^2	0.1821	0.2336
Log-likelihood	-135.21	-85.73
Correctly classified observations	74.9%	85.2%

Standard errors in parenthesis (), *denotes significance at 10% level, **denotes significance at 5% level.

Table 2.4. The impact of cooperative membership on commercialisation (PSM), ECX sites, 2005.

	Matching technique	
	Kernel	Nearest neighbour
[All Coop Members] – [Individual Farmers]	**0.06** (0.06)	**0.00** (0.08)
	78 members	*78 members*
	162 individuals	*48 individuals*
[Marketing Coop Members] – [Individual Farmers]	**0.14** (0.08)*	**0.21** (0.12)*
	42 members	*42 members*
	153 individuals	*25 individuals*

ATT in bold, standard errors in parenthesis (), *number of observations per group in italics.*
*denotes significance at 10% level, **denotes significance at 5% level.

markets, unless these cooperatives involve activites for collective output marketing. The robustness of these findings is supported by the fact that the Tobit regressions presented in Table 2.5 report very similar results.

Potential reasons underlying the positive impact of marketing cooperatives on smallholders' commercialisation involve the implicit cost-saving and risk-sharing devices of collective marketing, as documented in numerous studies (Nourse, 1945; Dulfer, 1974; Bonin *et al.*, 1993; Helmberger and Hoos, 1995; Munckner, 1988). On the other hand, potential reasons underlying the insignificant impact of all cooperatives on farm output commercialisation involve the 'defensive' attitude, related to prevalent rent-seeking behaviour, typical of non-

Table 2.5. The impact of coop-membership on commercialisation (Tobit), ECX sites, 2005.

Dependent variable: Commercialisation	All coops	Marketing coops
Cooperative membership	0.14 (0.10)	0.26 (0.10)**
Fixed arable land (hectares)	0.05 (0.05)	0.02 (0.05)
{Fixed arable land}^{2}	-0.00 (0.00)	0.00 (0.00)
Household size (no. of members)	0.05 (0.03)**	0.07 (0.03)**
{Household size}^{2}	-0.17 (0.38)	-0.16 (0.39)
Dependency ratio (children/adults)	-0.12 (0.05)**	-0.15 (0.05)**
Education of household head (years)	0.05 (0.03)*	0.03 (0.03)
{Education of household head}^{2}	-0.00 (0.00)**	-0.00 (0.00)
Age of household head (years)	0.01 (0.02)	0.01 (0.02)
{Age of household head}^{2}	-0.00 (0.00)	-0.00 (0.00)
Dummy for male household head	0.01 (0.13)	0.03 (0.13)
Distance to nearest market (min on foot)	-0.03 (0.00)**	-0.03 (0.00)**
{Distance to nearest market}^{2}	0.00 (0.00)**	0.00 (0.00)**
No. of observations	279	243
Pseudo R^2	0.1468	0.1579
Log-likelihood	-234.72	-204.17
Left censored observations	104	93
Uncensored observations	132	112
Right-censored observations	43	38

Standard errors in parenthesis (), *denotes significance at 10% level, **denotes significance at 5% level.

marketing cooperatives.[12] While the main role of marketing cooperatives is to reduce transaction costs and improve bargaining power of smallholders vis-à-vis the market, the main scope of defensive cooperatives is to reduce transactions costs and bargaining power of smallholders vis-à-vis the state and NGOs. As a result defensive cooperatives are major channels for aid, providing no incentives to small-scale entrepreunership, while marketing cooperatives represent a major channel for agricultural output flow towards new and more profitable markets. It is also interesting to note that the findings of the PSM and the Tobit regression analysis partly contradict with the results of the t-tests for the comparison of means presented in Table 2.2. Using t-tests, the commercialisation index is found to be significantly higher among cooperative members, regardless of the type of cooperatives. In other words, the results of naïve comparisons may lead to wrong conclusions, and confirm the need of using methods that control for diffusion effects and self selection, and take control variables into account.

2.6 Conclusions and implications

Ethiopia is witnessing the establishment of its first commodity exchange (ECX), for grains, pulses, oil seeds and coffee. The ECX represents a great opportunity to boost agri-commodity commercialisation and thus to promote agricultural growth and alleviate longstanding poverty in rural Ethiopia. However, unless access barriers to agri-commodity exchange are reduced, smallholder farmers may remain once again at the margin of economic development. For these reasons, the Ethiopian government is promoting the formation of smallholders' cooperatives all over rural Ethiopia, as well as their close interaction with the ECX. This study analyses the impact of membership of different types of cooperatives on agri-commodity commercialisation in major production areas targeted by the ECX.

Using re-sampling techniques, propensity score matching, and Tobit regression analysis, it is found that membership of a cooperative does not have a significant impact on the degree of coomercialisation. For members of marketing cooperatives, however, the degree of coomercialisation is between 14 and 26 percent higher than that of farmers who do not belong to a cooperative. Unless collective action involves collective marketing, agricultural cooperatives may not help smallholders to access agri-commodity markets and benefit from the ECX. The robustness of these findings is supported by the fact that two separate estimation techniques (Tobit regression and PSM) yield similar result.

In line with our conclusions, Bernard et al. (2008) find that membership of Ethiopian cooperatives has no impact on grain commercialisation on average. They explain their result from the heterogeneity in the responses of cooperative members to the cost-saving and risk-sharing advantages obtained through collective action. In particular, when facing a price increase smaller farmers (i.e. farmers with less land) tend to reduce the fraction of

---

[12] We also estimated the impact of membership of non-marketing cooperatives on commercialisation, using the same methodologies applied for all cooperatives and marketing cooperatives. The results indicate that membership of non-marketing cooperatives does not significantly affect the degree of commercialisation of farm households.

output marketed (i.e. sell less and consume more), whereas larger farmers tend to boost their commercialisation. Our analysis adds to the conclusions of Bernard *et al.* (2008) in the sense that it takes into account major organisational differences between cooperatives. In particular the distinction made between marketing and defensive cooperative allows us to advance and test the hypothesis that beyond heterogeneity in members' behaviour, heterogeneity in organisational behaviour plays also an important role in determining the impact of collective action on farm level commercialisation.

This study supports the remark made by Cook in a presentation at the Institut de Recherche Agronomique (Montpellier, 2007) *'a cooperative is not a cooperative is not a cooperative is not a...'*, meaning that cooperatives are not all the same, and different types of cooperatives serve different purposes. To pursue commercial purposes, Ethiopian agricultural cooperatives should serve their farmers not only by procuring cheap farm inputs (defensive organisational attitude), but also through collecting and selling farm output (offensive organisational attitude). We define the latter services as 'collective marketing' and we indicate it as the key activity for smallholders to gain access to agri-commodity markets such as the ECX. However, we observe that almost half of Ethiopian rural cooperatives do not engage in collective marketing but rather serve as a shield to protect semi-subsistence farming systems from market competition. In order to achieve the objective set by national public policy *'.. it has become necessary to establish cooperative societies .. and to enable cooperative societies to actively participate in the free market system ..'* (FDRE, 1998)[13], it is therefore crucial to improve the focus of development efforts towards the promotion of marketing cooperatives rather than any type of cooperatives, especially within the major agri-commodity production sites that are primary targets of the ECX. However, because large farmers are more likely to become members of (marketing) cooperative than small farmers, the extent to which promotion of marketing cooperatives contributes to poverty reduction is not yet clear. There is an urgent need for more empirical research on this issue. Further research is also needed to identify key factors behind the choice to form either a marketing or a defensive cooperative, as well as governance and managerial practices to maximise the sustainability of collective marketing activities over time.

---

[13] Later re-affirmed in the Sustainable Development and Poverty Reduction Program (FDRE, 2002), in the Plan for Accelerated and Sustainable Development to End Poverty (FDRE, 2005), and in the development plan (2006-2010) of the Federal Cooperative Commission;

Chapter 3. The life cycle of agricultural cooperatives: implications for management and governance in Ethiopia

Abstract

Commercialisation through cooperatives, i.e. collective marketing, has the potential to reduce transaction costs and improve bargaining power of farmers vis-à-vis the market. The objective of this study is to evaluate the probability for an Ethiopian agri-cooperative to engage in collective marketing activities over time, given market and governance characteristics. Using a sample of 200 agricultural cooperatives from the Ethiopian Highlands, the analysis suggests that collective marketing faces cyclical challenges related to increasing competition. Empirical results also suggest that among Ethiopian cooperatives, those established in the northern regions of Tigray and Amhara, and/or upon the voluntary initiative of farmers, embark on more sustainable collective marketing activities over time. The study concludes with implications for policy and further research.

3.1 Introduction

Historical experiences in industrial countries indicate that a key to advance agroindustrialisation is to simultaneously generate technological and institutional innovation (Hayami and Otsuka, 1992). As demonstrated by the limited benefits brought by the 'Green Revolution' in Africa, improved production, processing and marketing technology is not sufficient to advance agroindustrialisation. Even if improved technology (e.g. improved livestock and seeds) was made available in Africa, a myriad of smallholder farmers could neither access nor sustain it, mainly because of missing markets (Fafchamps, 2005; Von Braun, 1995).

Perhaps markets are missing in Africa because of the scramble of indigenous socio-economic institutions that occurred during colonial history (Bertocchi and Canova, 2002), or because of the dependency created from unsustainable foreign institutions, like NGOs, in postcolonial times (Keyzer and Wesenbeeck, 2007). For these reasons, scholars and policy makers are increasingly looking for ways to promote the development of indigenous, community driven market institutions (Binswanger, 2006), in an effort to realign institutional with technological development and advance agroindustrialisation.

This study focuses on Ethiopian agricultural cooperatives, which are one example of traditional market institutions. Although forms of rural cooperation in Ethiopia can be traced back in time almost to the origin of agriculture (7000-4000 B.C. according to Ehret, 1979), the institutionalisation of agri-cooperatives came only with the Derg and its communist regime (1974-1991). With the downfall of the Derg regime and its highly centralised governance, agricultural cooperatives entered a period of uncertainty during which many of them collapsed throughout the country. Before the downfall of the Derg, agricultural cooperatives became a major target of political propaganda by the government and the opposition, fuelling internal corruption and conflicts.

Nonetheless, since 1994 agri-cooperatives began to re-emerge, strongly promoted and supported by policy reforms envisaging a return to cooperatives as a way to improve the participation of smallholder farmers in the emerging national market (FDRE, 1994, 1998, 2002, 2005). According to Bernard et al. (2008) the share of kebeles with cooperatives went up from 10 percent in 1991 to nearly 35 percent in 2006.[14] In 2002 cooperative governance was further reinforced by the establishment of the Federal Cooperative Commission (FCC), a governmental body with the ambitious mandate to establish one cooperative per kebele to 70 percent of the national kebeles by 2010 (FCC, 2006). Although agricultural cooperatives have been growing rapidly in Ethiopia, and are expected to grow further, their contribution to improve agricultural commercialisation appears still negligible (Bernard et al., 2008). In Ethiopia, most agricultural cooperatives serve farmers to procure improved and subsidised farming inputs from the state (see Spielman et al., 2008), but only some of these cooperatives assist farmers to improve output marketing.

Commercialisation through cooperatives, i.e. collective marketing, has the potential to reduce the transaction costs and improve the bargaining power of farmers vis-à-vis the market (Munckner, 1998; Helmberger and Hoos, 1995; Bonin et al., 1993; Dulfer, 1974; Nourse, 1945; see also Chapter Two). However, in most Ethiopian cooperatives agricultural commercialisation takes place outside the cooperative system, depending exclusively on individual entrepreneurship and resources (Bernard et al., 2008).

A widespread opinion is that public interventions to promote the formation of rural cooperatives are often too invasive, creating collective dependency rather than collective entrepreneurship. Similar concerns are reported from many other developing countries, where cooperatives appear to be often used as instruments to implement policies designed without consulting them, in order to fulfil the agenda of the donors (World Bank, 2007: 156). Top-down interventions tend to attract opportunistic and subsistence farmers, eager to extract subsidies rather than embark in marketing activities. Cooperatives founded on the spontaneous initiative of farmers are instead more likely to aim for commercial objectives.

---

[14] In Ethiopia a kebele is the smallest administrative units, below the municipality-district level.

Cooperatives may fail to provide marketing services to their members also because they typically operate in the context of rural communities where they are subject to norms and values of social inclusion and solidarity (World Bank, 2007: 155). This may clash with the requirements of professional, business-oriented organisations that must help members compete in the marketplace. In the name of social inclusion and solidarity cooperatives can be pressed to include and cross-subsidise poorer-performing farmers at the expense of better performers, thereby weakening rewards for efficiency and innovation.

Another reason of collective marketing failure can be related to poor managerial capacity. In developing countries, agricultural cooperatives are usually managed by village elders or elites, who often lack the necessary skills and resources to sustain collective business over time (World Bank, 2007: 156). According to Putterman (1985), and Cook and Chambers (2007), collective marketing faces cyclical challenges. The marketing cycle is characterised by an initial stage with high turnover, followed by a reduction in sales due to increasing competition. Subsequently, cooperatives need to re-adjust their strategic behaviour to keep competing in the marketplace. However, smallholder cooperatives in developing countries may easily fail to do so, justifying external interventions. Governments and NGOs have an important role to play in supporting capacity building towards sustainable cooperative business management (World Bank, 2007: 156). However, external support to cooperative management has often resulted in political interferences on members' decisions, leading to internal corruption and conflicts (World Bank, 2007: 156).

The objective of this study is to evaluate the probability for an Ethiopian agri-cooperative to engage in collective marketing activities over time, given (external) market and governance conditions. To do so section two elaborates further on the collective marketing framework. Section three presents the data available and the characteristics of the sample. Section four defines the empirical model used to interpret the data. Section five discusses the findings, and section six draws conclusions and implications.

3.2 Analytical framework

Twentieth century economic scholars (see Staatz, 1987; Sexton, 1986; Sexton and Iskow, 1988) have generally agreed that agricultural cooperative business emerges because of conducive public policy, in markets affected by asymmetric information and monopsony (or monopoly), or oligopsony (or oligopoly) power. The existence of any one of these conditions leads to the consideration of collective action as a means to facilitate agri-business activities. By contrast, when public support is absent, and/or markets are missing or highly competitive, subsistence (autarkic) farming systems or investor owned firms are more likely to emerge.

Cooperatives in developing countries frequently face life cycle phenomena related to changes in their internal organisation and external market position (Putterman, 1985). Figure 3.1 shows the business cycle of the average US agri-cooperative, as reported by Cook and Chambers

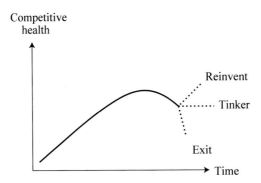

Figure 3.1. Cooperative life cycle. Source: Cook and Chambers, USA, 2007.

(2007). At an early stage cooperatives manage to procure and sell at lower prices than market competitors can do. As a result, cooperatives enter a period of growth and glory. However, while cooperative members tend to over-celebrate their achievements, market competitors begin to modify their strategic behaviour, and the competitive advantage of cooperatives begins to diminish. When cooperatives realise the pressure of increasing market competition they also realise the complexities they have to face to upgrade their business performance. While some members might be willing to invest in the common cause, others might not.

As stated by Olson (1965): *'...unless the number of individuals in a group is quite small or unless there is coercion or some special device to make individuals act in their common interest, rational, self-interested individuals will not act to achieve their common or group interest'*. Due to fading competitiveness and diverging preferences over time, disagreements and conflicts arise within cooperatives, undermining the stability of the coalition (Sexton, 1986; Staatz, 1987), and promoting the desertion of most progressive members (Barham and Childress, 1992; Cook, 1995; Karantininis and Zago, 2001). According to Barham and Childress (1992), the desertion of cooperative members can be considered as a natural adjustment process to reduce internal heterogeneity of preferences.

At some point cooperatives need to confront the decision to exit the market or to re-adjust (tinker or reinvent) their structure and conduct, and enter a new business cycle (Figure 3.2). The tinker option can involve investments made with external funds generated through srategic alliances with firms or other cooperatives. Alternative solutions can involve proportionality strategy of internally generated equity capital, such as base capital plans, proportional voting, narrowing product scopes, pooling on a business unit basis, and capital acquisition on a business unit basis. The reinvent option considered is that of shifting to a more radical or new form of cooperative such as a 'new generation cooperative' (see Sykuta and Cook, 2000). This new structure involves shareholding as a mechanism to generate equity capital, in addition to members' patronage (i.e. percentage of members' revenue retained by the cooperative). Where shares are irredeemable, tradeable and appreciable and members are required to

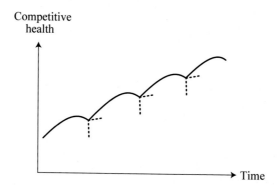

Figure 3.2. Consecutive cooperative life cycles. Source: Cook and Chambers, USA, 2007.

purchase them on the basis of expected patronage, so that patronage and shareholding are proportionately aligned.

According to Cook and Chambers (2007) the life cycles of US cooperatives describe an unpward trend (Figure 3.2), meaning that agricultural cooperatives are a sustainable form of business organisation in the US. However, different scenarios can be hypothesised for different countries. In particular in developing countries, markets and governance regimes are expected to be less favourable to cooperative business development. In developing countries, agricultural cooperatives are also village-level organisations that rely on limited managerial capacity. Kotler (1995) argues that organisational adjustments can be difficult in the absence of a common vision and a strong management. Pagano (1993) suggests that timely institutional reforms can be difficult to enforce when options for attracting the necessary venture capital are limited. For these reasons cooperatives in developing countries may tend to delay the organisational adjustments needed to keep competing in the marketplace. Due to adverse external conditions (missing markets and invasive governance), and managerial procrastination, agri-cooperative business in developing countries is expected to be less sustainable than in the US.

3.3 Data

The data used in this study were collected in Ethiopia through direct interviews with the management committees of 206 agricultural cooperatives. Sample sites (see also Figure 3.3) include the four regions of Tigray, Amhara, Oromia and SNNP, covering mainly the Ethiopian Highlands, which are generally characterised by favourable agro-ecological conditions. The sample includes 13 *woredas* (alike municipalities or district) per region, and four agricultural cooperatives per *woreda*.[15] The sample does not claim representativeness of the national agri-cooperative system. The survey was conducted between May and July 2006, and each cooperative

---

[15] In two *woredas* we were able to survey only three cooperatives.

Sample sites

Figure 3.3. Sample sites, Ethiopian Highlands, 2006. Source: IFPRI-ESSP.

was surveyed once on the basis of a structured questionnaire. The latter was designed with the intention to capture the heterogeneity in cooperative structure and conduct.

Within our sample, 62 percent of the cooperatives were established during the previous (Derg) regime (1974-1991), while the others emerged between 1993 and 2006 under the current government (post 1991). During the Derg, output marketing by cooperatives was directly organised and controlled by the state. The structural adjustments that followed the military coup and the fall of the Derg regime had profound impacts on existing agricultural cooperatives. As governance and markets were reformed, cooperatives had to re-organise to legitimate the continuation of their activities. Some were unable to do so and collapsed at the end of or immediately after the Derg regime. Others engaged in internal restructuring and re-institutionalisation. For these reasons, the data used in this study describe the establishment or re-establishment (for cooperatives originally established during the Derg regime), and the development of collective marketing in the period between 1991 and 2006 (post Derg).

Consequently, the age of cooperatives, measured from establishment (for cooperatives founded after 1991) or re-establishment (for cooperatives founded during the Derg and re-established after 1991) until 2006, ranges from a minimum of one to a maximum of 14 years, with an average of 12 years. 52 percent of the cooperatives were established or re-established on the initiative of farmers, as opposed to external initiatives by governmental or non-governmental

Table 3.1. Differences across cooperatives, Ethiopian Highlands, 2006.

Numer of Obs. 206	Coops that are not engaged in collective marketing	Coops that are engaged in collective marketing
Coops established on farmers' initiative (dummy)	0.31 (0.47)**	0.68 (0.47)**
Coops with 1st chairman appointed by the government (dummy)	0.37 (0.49)*	0.48 (0.50)*
Coops in Tigray (dummy)	0.17 (0.38)**	0.35 (0.48)**
Coops in Amhara (dummy)	0.15 (0.36)**	0.35 (0.48)**
Coops in Oromia (dummy)	0.41 (0.49)**	0.15 (0.36)**
Coops in SNNP (dummy)	0.27 (0.45)**	0.15 (0.36)**

Standard deviation in parethesis ().
* denotes significant difference between the two groups at 5 percent level.
** denotes significant difference between the two groups at 10 percent level.

organisations. The number of founding members can vary widely in Ethiopian cooperatives (10-3,000), and on average amounts to 600 farmers. In 2006, the average cooperative counted 884 members. The average growth in number of members from establishment to 2006 is estimated at 190 percent. In 44 percent of the cooperatives the initial chairman was appointed by the government. 60 percent of the cooperatives engaged in collective marketing, at least once, in the year before the survey. In our sample, agricultural marketing through cooperatives involves primarily cereals, such as teff (21 percent of the cooperatives), maize (18 percent) and wheat (9 percent), or coffee (16 percent).

In Table 3.1 we compare differences in the establishment of cooperatives that engaged in collective marketing between 2005-2006 and those that did not. Table 3.1 suggests that marketing cooperatives are mainly found in Tigray and Amhara regions. Table 3.1 suggests also that cooperatives established upon members' initiative, with an initial chairman appointed by the government, are more likely to engage in collective marketing. However, the analysis presented in Table 3.1 could be affected by selection bias due to the presence of cooperatives that did not engage in collective marketing because they were recently established and did not have sufficient time to set up marketing services.

Figure 3.4 shows that the probability to be engaged in collective marketing in 2005-2006 decreases with the age of the cooperatives, describing a concave curve.[16] The probability

[16] The probability for a cooperative to be engaged in collective marketing activities, given its age, is calculated using Locally Weighted Least Squares (or lowess smooth) technique (default in STATA).

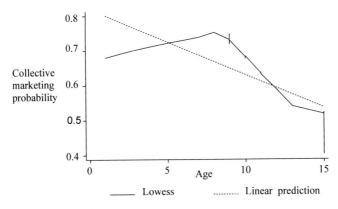

Figure 3.4. Life cycle of an average cooperative, Ethiopian Highlands, 2006.

increases during the first eight years of cooperatives' life, and then begin to decrease at a faster pace reaching a marketing probability that is below the initial level. The downward slope of the life cycle suggests that the average Ethiopian cooperative is an unsustainable form of marketing organisation over time. However, the scenario presented in Figure 3.4 could also be affected by selection bias since the sample used does not aim at national representativeness. Moreover Figure 3.4 neglects potential differences across cooperatives. In particular, there might be a minority of cooperatives that do engage in sustainable marketing activities. The following part of the analysis needs to identify these succesful stories, if they exist, as well as their market and governance framework.

3.4 Empirical model

The empirical model presented in this section aims at measuring the probability for an Ethiopian agri-cooperative to engage in output marketing activities over time, given the market and governance environment in which it operates. To do so, we estimate the following probit model:

$$y_i = \beta_0 + \beta_1 (m_i * x_i) + \beta_2 (m_i * x_i^2) + \beta_3 (m_i * x_i^3) + \beta_4 (g_i * x_i) + \beta_5 (g_i * x_i^2) + \beta_6 (g_i * x_i^3) + \beta_7 l_i + e_i \quad (3.1)$$

where the dependent variable, y, is equal to one when a cooperative i engaged in output marketing activities during 2005-2006, and equal to zero when it did not. In order to capture the cyclical evolutions of cooperative business, the independent variables in Equation 3.1 include cooperative age, x, as well as its squared value, x^2, and cubic term, x^3. Since Ethiopian cooperative business evolves in cycles (see Figure 3.3) these variables are expected to explain y, with x^2 showing opposite sign in respect to x and x^3.

In order to distinguish the effect of different markets and governance regimes on the cyclical evolution of Ethiopian agri-cooperatives, x, x^2, and x^3 are interacted with two indicators: (1)

a dummy, *m*, for cooperatives established on farmers' initiative (*m* equal to one), as opposed to cooperatives originated from top-down initerventions by the government or NGOs (*m* equal to zero); and (2) a dummy, *g*, for cooperatives whose initial chairman was appointed by the government (*g* equal to one), as opposed to cooperatives with an initial chairman SNN chosen by the farmers (*g* equal to zero).

As discussed in section two, cooperatives founded on the initiative of a small group of members, under the conducive support of the state, are more likely to sustain marketing activities over time. For this reason farmers' initiative, *m*, is expected to have a positive influence on collective marketing in 2005-2006, *y*. Part of the literature discussed in section two suggests that governmental interference, *g*, has a negative impact on collective marketing. However, when cooperatives are formed by poorly educated smallholders the intervention of the government could also be necessary to promote collective marketing. The empirical model (Equation 3.1) includes also a set of three dummies, *l*, indicating the region in which a cooperative *i* operates (Tigray, Amhara, Oromia, or SNNP)[17]. Ethiopia is a Federated Republic, in which regional governance is semi-autonomous, and Amhara and Tigray regions have a longer history of trade and are also more advanced in terms of infrastructures, urbanisation, and institutions, compared to the rest of the country. Regional differences reflect the fact that Amhara represented the ethnic elite during the longstanding empire (1930-1974) of Haile Sellaise (himself an Amhara), while Tigray is the homeland of the current ruling party.

The empirical model proposed (Equation 3.1) could suffer from econometric problems inherent to the use of cross-section data, and these should be addressed before interpreting the results. In most cases, when econometric models are based on data collected at one point in time, as in this case, it is difficult to ascertain that right hand side variables cause variations in the left hand side variable rather than the other way around (endogeneity). However, causality does not seem to be a problem in this model since age of (existing) cooperatives, and lagged variables (referring to cooperatives' establishment) are interacted in the model. An additional concern relates to the use of cross section data is heteroskedasticity, here controlled by estimating the model with robust standard errors.[18]

3.5 Results

Empirical findings are summarised in Table 3.2. Results suggest that the regions of Tigray and Amhara offer better environments indeed for agricultural cooperatives to embark in collective marketing activities. Cooperatives in these two regions have 23-27 percent more probability to engage in collective marketing than in the other two regions (SNNP and Oromia). These findings are supported by the frequent complaints (sometimes degenerating into violent acts) of southern populations (from SNNP, Oromia, Gambela and Somali

---

[17] The regions covered by the survey are four.

[18] Heteroskedasticity occurs when the variance of the random error term is not constant across observations.

Table 3.2. Heterogeneity in cooperative life cycles (Probit), Ethiopian Highlands, 2006.

Dependent variable: dummy for cooperatives that engaged in collective marketing in the last year (2005-06)	Probit estimation	Marginal effects
Coops established on farmers' initiative		
Coop age	2.40 (0.58)**	0.89 (0.20)**
Coop age^2	-0.44 (0.10)**	-0.16 (0.04)**
Coop age^3	0.02 (0.00)**	0.01 (0.00)**
Coops with 1st chairman from the government		
Coop age	-0.24 (0.33)	-0.09 (0.12)
Coop age^2	0.07 (0.06)	0.03 (0.02)
Coop age^3	-0.00 (0.00)	-0.00 (0.00)
Spatial effects		
Coops in Tigray (dummy)	0.61 (0.34)*	0.21 (0.10)**
Coops in Amhara (dummy)	0.72 (0.30)**	0.24 (0.09)**
Coops in Oromia (dummy)	-0.10 (0.31)	-0.04 (0.11)

Number of obs. = 201	Correctly classified obs. = 75.6%
Log pseudolikelihood = -103.68	Pseudo R^2 =0.2349

Standard error in parenthesis (), *denotes significance at 10% level, **denotes significance at 5% level.

regions) about political clientelism, in favour of Tigray and Amhara regions. Table 3.2 shows also that farmers' initiative is significant in explaining the probability for a cooperative to be engaged in collective marketing in 2005-2006, given cyclical evolutionary patterns. By contrast, governmental interference in cooperative management is insignificant in explaining collective marketing probability.

The relationship between farmers/external initiative and collective marketing, over time, is depicted in Figure 3.5. It is clear that cooperatives established upon farmers' initiative are a more sustainable form of business than cooperatives established on the basis of top-down initiatives (by either the government or NGOs). This finding is largely supported in development and agri-business literature (see section two), which generally recognises the voluntary and active participation of farmers as key indicator of commitment to collective entrepreunership. The literature appears to be fairly divided on the issue of public interference in cooperative management (see section two). Our empirical results suggests that governmental interference in cooperative management has no significant impact in promoting collective marketing acivities.

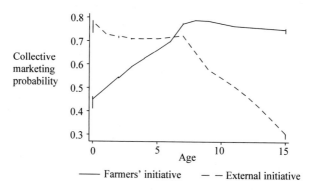

Figure 3.5. Life cycle of bottom-up and top-down cooperatives, Ethiopian Highlands, 2006.

3.6 Conclusions and implications

Throughout history, Ethiopian rural households have formed various forms of associations (or cooperatives) to solve their socio-economic problems. Today, agricultural cooperatives are seen as an institutional solution to support the livelihood and commercialisation of Ethiopian farmers. However, this study suggests that agricultural commercialisation through cooperatives faces cyclical challenges, and that Ethiopian cooperative cannot always sustain collective marketing activities over time.

Collective marketing appears to be more sustainable in Tigray and Amhara regions where market and/or governance conditions are more favourable than in the southern Ethiopian regions. Furthermore, collective marketing activities appear to be more sustainable in cooperatives established on the voluntary initiative of farmers, than in cooperatives formed by top-down interventions (by the government or NGOs). External interventions increase the probability for a cooperative to embark on collective marketing at an initial stage. However, collective competitiveness decreases rapidly in cooperatives formed by the government and NGOs. Cooperatives founded on the voluntary initiative of farmers are instead less likely to engage in collective marketing at an early stage, but they are more likely to sustain these business activities over time. This study shows also that the direct interference of the government in cooperative management brings no clear benefits to collective competitiveness.

For these reasons, public support to agricultural cooperatives should avoid direct interference with establishment and management processes, but it should rather focus on building managerial capacity, so as to prepare cooperative members to confront the cyclical challenges coming from the marketplace. Further research is needed to identify good managerial practices to be applied by different typologies of cooperatives in different market environments.

Chapter 4. Evolution and challenges of dairy supply chains: evidence from supermarkets, industries and consumers in Addis Ababa

Abstract

Although the livelihood of a major share of the Ethiopian rural population depends on the production of staple cereals, the demand pressure for high value and perishable products like dairy is expected to grow rapidly in the coming years. Using data collected among 200 urban households, this study presents a detailed analysis of supermarket-led dairy chains in Addis Ababa, with the purpose to examine the recent evolution of the dairy market and the challenges posed to Ethiopian producers. First, we estimate that supermarkets account for 11 percent of the total sales of dairy products in Addis Ababa. The vast majority of the dairy products sold by supermarkets are procured from modern industrial manufacturers who accounts for 17 percent market share. Second, econometric analysis suggests that selling of processed dairy products through supermarkets will continue to grow as urbanisation and incomes increase. Finally, we observe that the degree of power concentration in the dairy processing industry, and the degree of spatial concentration of supermarket outlets are both very high. The study concludes with some implications for public policy and further research.

4.1 Introduction

Population growth, urbanisation and income growth are occasioning a massive increase in demand for dairy and other food products of animal origin in the Middle East, North and sub-Saharan Africa (Delgado *et al.*, 1999). FAO-IFPRI-ILRI projections indicate that dairy consumption is estimated to grow by an average 3.8 percent per year in the sub-Saharan region, and by three percent in North Africa and the Middle East.[19]

Although the livelihood of a major share of the Ethiopian rural population depends on the production of staple cereals, the country has favourable agro-ecological conditions for milk production, and great potential to meet regional milk demand. Still, the production system remains dominated by a myriad of smallholder farmers who produce mainly for home consumption or sell to neighbouring households (Ahmed *et al.*, 2003). Staal (2006a) reports

---

[19] The projected growth rate of dairy consumption for sub-Saharan Africa is the second largest in the world after India (4.1%) (Delgado *et al.*, 1999).

that 78 percent of the milk produced in Ethiopia is consumed by producer households. In Ethiopia per capita milk production is estimated at 41.6 litres per year by Taffesse et al. (2006), and 20 litres per year by FAO (FAOSTAT, 2006[20]). Although these estimates differ considerably, they are much lower than the figure reported for Kenya (90-100 litres per year; FAOSTAT, 2005[21]).

Milk is a perishable product, and smallholder farmers willing to improve milk production and commercialisation need to link up with dairy manufacturing and retailing chains, so as to reach farther consumers (Staal et al., 2001). However, modern supply chains pose important challenges to smallholder farmers (World Bank, 2007: 156). A dairy supply chain can be disentangled into downstream operators, such as manufactures and retailers, and upstream operators such as farmers. The role of downstream operators is to systematically address, if not anticipate, evolutions in consumer preference; whereas the role of upstream agents is to keep up with the increasingly stringent specifications of manufacturers and retailers (see Weaver and Kim, 2001).

Supermarkets have recently emerged in many parts of Africa triggering profound changes in supply chains (see Neven et al., 2006; and D'Haese and Van Huylenbroeck, 2005; among others). According to the World Bank (2007: 126), supermarkets emerge in large cities, and then spread to smaller towns. First they target the upper-income consumers, then the middle class and later also the urban poor. The supply of supermarkets is dominated by processed food products with extended shelf life. Berdegue et al. (2005) for Central America, Dries et al. (2004) for Europe, and Wheatherspoon and Reardon (2003) for Africa, show that supermarkets brings major changes in the procurement systems for canned, dry and packaged food, especially for meat and dairy.

As documented by Humphrey (2007), Reardon et al. (2007), Trail (2006), Berdegué et al. (2005), Reardon et al. (2005), D'Haese and Van Huylenbroeck (2005), and Dries et al. (2004), the rise of supermarkets is driven by industrialisation and changes in consumer behaviour, associated with increasing urbanisation and purchasing power. However, the expansion of food outlets and advances in industrial technology tend to concentrate market power in the hands of industrial-retail oligopolies (see Eagleton, 2006; Weatherspoon and Reardon, 2003; Kaplinsky and Morris, 2001). Hence, the rise of supermarket-led supply chains is a consolidated global trend, associated with both societal opportunities and discontents.

Are supermarket-led chains also emerging in Ethiopia? What is their share of the market? Are they expected to keep emerging and growing in the near future? And what are the challenges they pose to the Ethiopian society? To answer these questions we present a detailed analysis of supermarket-led dairy chains in Addis Ababa. The analysis is organised as follows. Section

---

[20] http://faostat.fao.org/site/567/DesktopDefault.aspx?PageID=567.

[21] http://faostat.fao.org/site/336/DesktopDefault.aspx?PageID=336.

two discusses the data that we collected for our analysis. Section three examines the rise of supermarkets in Addis Ababa and the type of dairy products that they sell. Section four analyses the evolution and current structure of the Ethiopian dairy manufacturing industry. Section five investigates the characteristics of dairy consumers in Addis Ababa and the reasons they give for buying or not buying dairy in supermarkets. Section six presents an econometric model that we use for analysing the factors driving outlet choice and the quantity of dairy products bought by urban households in Addis Ababa. Section seven discusses the empirical findings, while section eight presents our conclusions and implications.

4.2 Data

This study builds on national, regional, and international secondary data, as well as on primary information collected from consumers in Addis Ababa. Sources of secondary data include (inter)national literature and development agencies, the Chamber of Commerce of Addis Ababa, the Ethiopian Central Statistical Agency and the Ethiopian Ministry of Trade and Industry. The data obtained from these sources is mainly used to examine dairy industries and supermarkets.

Sources of primary data include 200 households from the urban area of Addis Ababa. Household data collection took place between March and May 2006 with the help of one enumerator (asking questions to the household member and translating the answers into English for the supervisor), one supervisor (cross-checking the consistency of the answers throughout the interview), and the use of a structured questionnaire. Households were selected using the following procedure. Among the 28 *woreda* (districts or sub-cities) of Addis Ababa we selected the one with the lowest, one with middle-low (randomly selected out of 13 potential candidates), one with middle-high (randomly selected out of 13 potential candidates), and the one with the highest income. The selection of the four *woreda* was based on the households' expenditure survey published in 2000 by the Central Statistical Authority (CSA). For each selected *woreda* we identified two neighbourhoods characterised by houses with average size and condition for that specific *woreda*. Interviews took place at every other house until 25 interviews were accomplished.

When a household refused to cooperate it was replaced by another. In order to minimise the number of non-cooperating households, interviews were conducted during lunch and dinner time, as well as on appointment. Overall, the response rate was very high (89 percent) except in the richest *woreda,* where the enumerators were rejected in 50 percent of the cases by compound guards. During the interviews, household heads were asked about their preferences, shopping frequency, and expenditures on different dairy products and retail outlets, as well as about the socio-economic characteristics of the household.

Clearly, the sampling frame that was used does not aim at producing results that are representative for the whole country. However, Addis Ababa is the biggest and most developed national market, and offers the largest choice of dairy products and retail outlets in Ethiopia. Therefore, we assume that changes taking place in Addis Ababa are likely to be a good indication of the changes taking place in smaller urban areas, the difference mostly being in scale and time lags.

Like most household samples, especially from developing countries, the household data available to this study is characterised by strengths and weaknesses that need to be taken into account. The main advantages and disadvantages of our sample are related to the stratification method adopted in selecting the households. This method emphasises household variability to the detriment of results' representativeness. While regression analysis is expected to benefit (in terms of goodness of fit) from the large variability across observations, descriptive analysis becomes more cumbersome as the results may not reflect the typical consumer of Addis Ababa, but rather the average consumer from the four selected *woreda*. To overcome this problem, the descriptive statistics presented in the following sections are calculated applying sampling weights, where appropriate. The weights that are used are reported in Table 4.1.

Table 4.1. Calculation of sampling weights, Addis Ababa, 2006. Source: calculated from census held by the Central Statistical Agency (CSA, 2000) in Addis Ababa.

Strata	Population (no. of households)	Sample (no. of households)	Sampling weights
Richest sub-city[1]	10,980	47	234
Middle-rich sub-cities (13)	188,989	50	3,780
Middle-poor sub-cities (13)	185,411	50	3,708
Poorest sub-city	14,231	50	285

[1] The richest subcity includes 47 households instead of 50 because three of the questionnaires filled in this subcity appeared to be incomplete and therefore discarded.

4.3 Supermarkets

In Ethiopia, over the past decade, supermarkets have emerged as an important agent of change in the urban food retailing systems.[22] This phenomenon reflects a well documented global trend. While in Germany, the US, UK, and France the share of supermarkets in domestic food retailing has reached 70-80 percent, in less developed countries supermarkets are less dominant but growing fast (Reardon, 2005). In India, although the share of supermarkets

[22] In this paper we use the definition of supermarkets given by Neven *et al.* (2006) 'self-service stores handling predominant food, drugs and household fast-moving goods (FMCG) with at least 150 m^2 of floor space'.

in food retail is only 5 percent, supermarkets are growing by 18-20% a year (The Economist, 2006). India is considered to be among the top three most attractive countries in the world for foreign direct investment in retail. China had no supermarkets in 1989, and the food retail sector was nearly completely controlled by the government. In 1990 the supermarket sector began to develop, and by 2003 had climbed to a 13 percent share in national food retail and 30 percent share of urban food retail, with 71 billion dollars of sales. The sector shows the fastest growth in the world, at 30-40 percent per year (Hu *et al.*, 2004). In Latin America, between 1990 and 2002, the share of supermarkets in domestic food retailing rose from roughly 15% to 55% (Reardon and Berdegue, 2002).

In the last decade, also the African urban markets have witnessed the rapid proliferation of supermarkets, particularly evident in Kenya and South Africa (Weatherspoon and Reardon, 2003). Supermarkets already have a 55 percent share of national food retail in South Africa, similar to the share in Argentina and Mexico (and not far behind the 70 percent in the United States; Weatherspoon and Reardon, 2003). South African 'Shoprite' is now the largest African retailer, with over 700 shops in 16 countries (The Economist, 2005). Supermarkets in Kenya have grown from a tiny niche market only seven years ago to 20% of urban food retail today (Neven *et al.*, 2006). As the driving factors behind the growth of supermarkets, i.e. growth of the urban population, increasing market liberalisation, competition and globalisation, are expected to continue over the next decade in Africa, the supermarket sector is expected to continue to grow (Neven *et al.*, 2006).

Within Ethiopia, the proliferation of supermarkets was particularly evident during the last five years in Addis Ababa, especially in and around the richest area of the city (Bole) where the number of supermarkets doubled. The very first supermarket (*Bambis*, which is still in business) was established in Addis Ababa at the end of the Imperial regime (1930-1974). At the moment there are 22 supermarkets registered with the chamber of commerce of Addis Ababa. Outside Addis Ababa the number of supermarkets remains very limited. Supermarkets in Ethiopia appear to be concentrated in a few wealthy neighbourhoods of the capital city, as observed also in other poor countries (World Bank 2007: 126).

Due to a lack of adequate information it is unfortunately not possible to quantify the share of supermarkets in the food retail sector of Addis Ababa. With respect to the dairy sector, however, our primary data shows that supermarkets sell on average 35,000 kg (in milk equivalent) of dairy products per day, accounting for an 11 percent share of the total dairy retailed in Addis Ababa in 2006. The role played by supermarkets is especially important for industrial dairy products (29 percent share), and less important for traditional dairy (2 percent share). Up to 91 percent of the dairy products retailed by supermarkets are industrially produced. Table 4.2 shows that as far as industrial dairy is concerned supermarket prices are not significantly different from the prices offered by other retailers.

Table 4.2. Dairy prices across outlets, Addis Ababa, 2006. Source: household data collected by the author.

Dairy outlet/product	Pasteurised milk (Birr/lt)	Powder milk (Birr/kg)
Supermarkets [65 obs.]	4.8 (1.2)	64.9 (30.4)
Other retailers [135 obs.]	4.7 (0.4)	56.4 (16.4)

Standard error in parentheses (), *denotes difference significant at 10% level, **denotes difference significant at 5% level.
1 Euro=11.5 Ethiopian Birr (in 2006).

4.4 Dairy industries

The dairy industry has a number of specific features that distinguish it from other agricultural industries (Schelhaas, 1999). Milk is highly perishable and produced on a daily basis. Milk requires timely management and involves high transportation and transaction costs. For these reasons, a major challenge of the global dairy industry is to extend the shelf life of products without affecting nutritional and sensorial attributes (Euromonitor International, 2004). Staal et al. (2001) show how the transformation from a supplier of 'traditional' dairy products, such as raw or fermented milk products, to a supplier of 'industrial' dairy products, such as standardised, sanitised (e.g. pasteurised, sterilised, etc.), and packed products, allows the dairy industry to access remote markets that previously were out of reach. In the specific case of the Addis Ababa's market, dairy products with extended shelf life can help to deal with the wide fluctuations in demand that are associated with the fasting practices of orthodox Christians.[23] Calculations based on our survey held among consumers in Addis Ababa show that daily consumption of milk and other dairy products decreases by almost 60 percent during fasting periods.

Nonetheless, the dairy market of Addis Ababa remains largely dominated by traditional dairy products that are produced within Ethiopia, such as raw milk, fermented butter mainly used for cooking (kebe'), and curdled skimmed products, including cottage cheese (ayb), and sour milk (ergo). Raw milk and kebe' are the dominant dairy products in Addis Ababa (Figure 4.1), like in the rest of the country (Ahmed et al., 2003). In Addis Ababa, the market share of the traditional dairy system is estimated at 75 percent (300,000 kg/day in milk equivalent; see Figure 4.2 and Ahmed et al., 2003).[24] The importance of traditional dairy in Addis Ababa is

---

[23] The calendar of the orthodox Christian church involves three prolonged fasting periods per year, and two fasting days every week (Wednesday and Friday), for a total of more than 200 days. During fasting days most orthodox Christians abstain from consuming products of animal origin. Orthodox Christians are grossly estimated at 40 percent of the national population. (Ahmed et al., 2003).

[24] One milk equivalent is equal to one litre of milk, 6.6 kilograms of butter, 4.4 kilograms of cheese, or 7.6 kilograms of milk powder (Jabbar et al., 2000).

Cooperation for competition

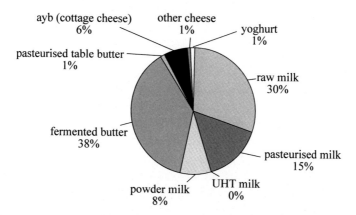

Figure 4.1. Dairy consumption breakdown by product, Addis Ababa, 2006.
Note: sampling weights are used in calculating these data. Source: household data collected by the author.

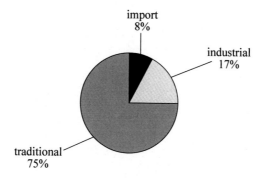

Figure 4.2. Dairy consumption breakdown by source, Addis Ababa, 2006.
Note: Sampling weights are used in calculating these data. Source: Household data collected by the author.

even higher than that reported by Staal (2006b) for Nairobi where the market share of the traditional dairy industry is estimated at 65 percent.

In Addis Ababa, industrial dairy products (with extended shelf life) include pasteurised, UHT and powder milk, and packaged (table) butter, cheeses (mainly *cheddar, provolone, mozzarella, gouda, feta, parmesan*), and yogurt. UHT and powder milk are usually imported from European and Arabic countries; packaged butter, yogurt and cheeses can be either imported or made in Ethiopia; while pasteurised milk is usually produced within 100 km from the capital. Dairy imports have an eight percent market share in Addis Ababa, while the dairy industry based in Ethiopia accounts for a 17 percent market share in Addis Ababa (Figure 4.2). Among industrial dairy products, pasteurised milk and, to a lesser extent, powder milk are the most widely consumed (Figure 4.1).

Price differences may play a role in explaining the limited penetration of industrial dairy products in the market of Addis Ababa. For instance, the price of one kilogram of traditional fermented butter is 18 percent lower than the price of one kilogram of packaged table butter (Table 4.3). Similarly, the price of one litre of pasteurised milk is 36 percent higher than the price of one litre of raw milk (Table 4.3). Staal (2006b) observes that Nairobi's consumers pay only 20 percent extra when they purchase pasteurised milk, instead of raw milk. Still, it is important to note that modern dairy products made in Ethiopia are consistently cheaper than imported dairy. As an example, one litre of pasteurised milk produced in Ethiopia and sold in Addis Ababa is 40 percent cheaper than one litre of milk reconstituted from imported milk powder (Table 4.3).

Table 4.3. Price differences across butter and milk products, Addis Ababa, 2006. Source: household data collected by the author.

	Birr/Kg
Butter	
Fermented butter [179 obs.]	36.4 (4.8)
Pasteurised butter [30 obs.]	44.3 (11.3)**
Milk	
Raw [89 obs.]	3.0 (0.6)
Pasteurised [137 obs.]	4.7 (0.5)**
Powder [37 obs.]	7.8 (2.9)**

Standard error in parentheses (), *denotes difference with traditional product price significant at 10% level. **denotes difference with traditional product price significant at 5% level.
1 Euro=11.5 Ethiopian Birr (in 2006).

Although its current performance may not be striking, the modern Ethiopian dairy industry has developed considerably in recent years. Dairy industrialisation began in 1979 in Ethiopia, when the Derg regime established the *Dairy Development Enterprise (DDE)* in Addis Ababa. The *DDE*, which is still operating under governmental ownership and control, had a capacity to process 60,000 litres of milk per day at its inception (Yigezu, 2000), but is currently processing only an average of 12,000 litres per day (as reported by Land o'Lakes in 2007).[25] With the downfall of the Derg regime in 1991, the private sector began to enter the national dairy industry as an important actor. Nowadays there are at least 12 private milk-processing plants operating in and around Addis Ababa (as reported by Land o'Lakes). One of these (Agro-Sebeta Industry, branding its products as 'Mama's milk') is processing an average of 28,000 litres of milk per day, outperforming the rival state-owned industry. The others are considerably smaller, processing an average of 1,000-3,000 litres per day.

The latter evidence suggests that indeed the private dairy industry is developing in Ethiopia, but also that competition is still at an infant stage within this sector. The national dairy industry appears largely dominated by a small number of firms. Figure 4.3 compares the concentration of the dairy industry in Addis Ababa with that of Nairobi in Kenya. In Addis Ababa, one firm (Agro-Sebeta) supplies almost 50 percent of the total modern dairy products available, against 25 percent supplied by the biggest manufacturer in Nairobi. And the two biggest Ethiopian manufacturers (Agro-Sebeta and DDE) supply almost 70 percent of the total modern dairy available in Addis Ababa, against 42 percent supplied by the two biggest firms in Nairobi. Computation of the Herfindahl concentration index reveals that the degree of concentration in the modern dairy industry supplying Addis Ababa is almost double that in Nairobi.[26]

---

[25] Land o'Lakes is a US dairy cooperative that devolves part of its capital and human resources to promote dairy businesses in less developed countries. During the preparation of this paper, the first author and Land o' Lakes representatives had several opportunities (formal workshops and informal meetings) to share information.

[26] The Herfindahl index (H) is a ratio showing the degree by which an industry is dominated by a small number of large firms or made up of many small firms (Encaoua and Jacquemin, 1980). It is defined as the sum of the squares of the market shares of each firm involved in the industry: $H = \sum f_n^2$, where f is the market share of firm n. A pure monopoly would take a value of one, while if all the firms in the industry had equal market shares the value would be $1/n$ (3 percent in Kenya and 8 percent in Ethiopia).

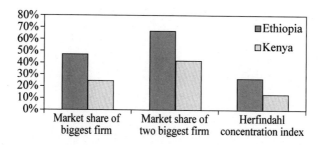

Figure 4.3. Degree of concentration in the dairy industry in Ethiopia and Kenya.
Source: calculated from information provided by Land o' Lakes (2007) for Ethiopia, and PKF Consulting Ltd, International Research Network (2005) for Kenya.

4.5 Dairy consumers

Although the urban population doubled (from 7 to 14 million) between 2000 and 2006 in Ethiopia, only 19 percent of the large Ethiopian population (73 million) was living in urban areas by 2006 (World Bank, 2007: 320, 334). However, supermarkets are found almost exclusively in Addis Ababa[27], which accounts for 4-5 million of the national urban dwellers (CSA, 2004). Table 4.4 shows descriptive statistics for some of the major variables on which we collected information in our survey among 200 households in Addis Ababa. As can be seen from the table, 15 percent of the consumers in Addis Ababa have purchased dairy products from supermarkets between spring 2005 and spring 2006. On average consumers in Addis Ababa purchased 11.4 kg (in milk equivalent) of dairy from supermarkets during that period.

The average household in Addis Ababa has 5 members, one of which is a child (below 14 years old). Adult females have on average 7.3 years of education and adult males have on average 7.5 years of education. In 75 percent of the households, adult females are responsible for dairy shopping. The fasting practices of the orthodox Christian religion are observed by 57 percent of the households. Only 12 percent of the households own a car and only 39 percent of them have a fridge. The average expenditure for a meal outside the household by a typical adult dweller is 5.5 Birr (approximately 0.5 Euro).

---

[27] This statement is based on field observations during four years of residency in Ethiopia, from 2003 to 2007.

Table 4.4. Household characteristics, Addis Ababa, 2006. Source: household data collected by the author.

Number of observations: 197 households	Mean	Std. deviation	Min	Max
Dairy shopping in supermarkets (dummy)	0.15	0.36	0	1
Dairy purchased from supermarkets (kg)	11.42	54.40	0	1135.32
Households owning a car (dummy)	0.12	0.33	0	1
Households owning a fridge (dummy)	0.39	0.49	0	1
Money spent for a meal outside the household (Birr)	5.46	12.56	0	150
Household size	5.26	2.34	1	20
Male education (years)	7.53	6.64	0	18
Female education (years)	7.29	5.30	0	18
Female responsible for dairy shopping (dummy)	0.75	0.44	0	1
Dependency ratio (no. of children/household size)	0.20	0.17	0	0.63
Household not observing fasting (dummy)	0.43	0.50	0	1

Note: sampling weights are used in calculating these data.

Figure 4.4 shows the answers to the question 'What are your reasons to purchase dairy from supermarkets or from traditional outlets' that we posed in the survey. Two reasons that are frequently given by consumers in Addis Ababa for buying dairy from traditional outlets is that supermarkets are too far from home or simply unknown. But the most frequently given answer refers to culture and habits. Many households in Addis Ababa have been purchasing from urban farmers, itinerant traders and wet markets for generations, and tend to consider supermarkets as 'fancy places for rich to shop'. Besides, some consumers mention that the taste and nutritional value of raw (whole) milk from urban farmers is superior than the taste and

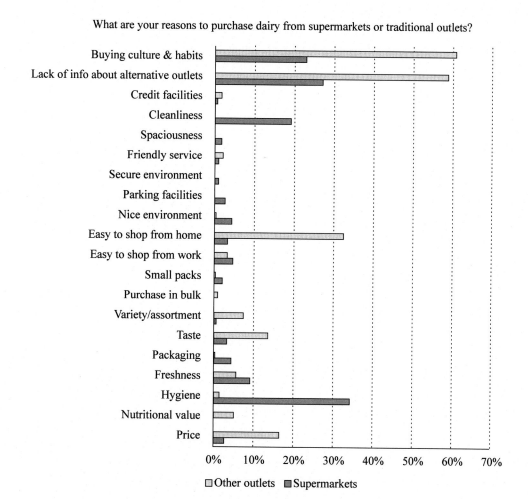

What are your reasons to purchase dairy from supermarkets or traditional outlets?

Figure 4.4. The opinion of dairy consumers, Addis Ababa, 2006.
Source: Household data collected by the author. Household preferences were obtained asking consumers in Addis Ababa to indicate the three most important reasons to shop, and not to shop, for dairy in supermarkets, given a list of 20 potential reasons.

nutritional value of pasteurised milk from supermarkets (usually standardised at 2.5 percent of fat content), or claim that itinerant traders and wet markets provide a larger variety and assortment of *kebe'* and *ayb,* which have a better taste than the modern alternatives found in supermarkets (respectively pasteurised table butter and cottage cheese). A surprisingly small share of the households (less than 20 percent) mentions the price difference as a reason for not buying dairy in supermarkets. Those who do buy dairy in supermarkets mention especially the superior hygiene of products and stores and the cleanliness, in addition to habits and lack of info on alternatives, as the reasons for doing so.

4.6 Model specification

Although it is clear that supermarket-led dairy chains are emerging in Addis Ababa, the determinants of its growth are only partly understood. Insights into such determinants can be used to identify target consumers and conceptualise expected evolution for supermarket-led dairy chains in the future. Mainstream theory (see section one) indicates urbanisation and purchasing power as major determinants underlying the emergence of supermarket-led supply chains. In Tanzania, Mdoe and Wiggings (1996) estimate the quantity of whole milk and reconstituted milk consumed per household as a function of income per capita, household size, number of children, education level and retail price. In Nigeria, Jansen (1992) explains the quantity of dairy consumed per person on the basis of income per capita, household size, number of children, education level, ethnic origin and location. Fuller *et al.* (2006) find that Chinese households with higher education levels buy dairy products more frequently from supermarkets and purchase more UHT milk.

Roux *et al.* (2000) explain that easy access is important in determining outlet choice by low income French consumers. Fuller *et al.* (2006), and Goldman and Hino (2005) argue that households with a fridge consume more dairy. Goldman and Hino (2005) and Neven *et al.* (2006) suggest that having a car facilitates shopping from supermarkets. Veeck and Veeck (2000) emphasise that women play an important role in the decision to shop in supermarkets. As supermarkets provide a larger assortment of food products, they attract households where all adult, including women, are working and have less time to shop (Goldman and Hino, 2005). Finally, Staal *et al.* (2006a) conclude from their study in Ethiopia that (orthodox Christian) fasting practices may indeed affect dairy purchasing behaviour.

We specify our model of dairy purchasing decisions in Addis Ababa based on these findings from the literature. With respect to the functional form to be used, Goldman and Hino (2005) estimate the probability that consumers shop in supermarkets by using a binary Probit model. Jansen (1992) and Mdoe and Wiggings (1996) estimate the volume of dairy products purchased with a simple linear regression. Neven *et al.* (2006) link such discrete and linear regressions by incorporating the predicted probability from the Probit model in the linear regression. By doing so, Neven *et al.* (2006) enlarge the spectrum of the analysis and control for selection bias in the linear regression (due to consumers that do not shop in supermarkets). We advance the latter methodology by applying a Heckman selection model (Heckman, 1979).[28]

The dependent variable in Equation 4.1, *smp*, is a dummy distinguishing between households that do (equal to one) and households that do not (equal to zero) shop for dairy in supermarkets. The dependent variable in Equation 4.2, *smq*, measures the total quantity (in milk equivalents) of dairy products purchased from supermarkets by a household i, during the last 12 months.

$$smp_i = \beta_0 + \beta_1 inc_i + \beta_2 hhsize_i + \beta_3 kids_i + \beta_4 medu_i + \beta_5 fedu_i + \beta_6 fast_i + \beta_7 shop_i + \beta_8 rw_i +$$
$$\beta_9 mrw_i + \beta_{10} mpw_i + \beta_{11}(kids * fast)_i + e_i \tag{4.1}$$

$$smq_i = \beta_0 + \beta_1 inc_i + \beta_2 hhsize_i + \beta_3 kids_i + \beta_4 medu_i + \beta_5 fedu_i + \beta_6 fast_i + \beta_7 shop_i +$$
$$\beta_8(kids * fast)_i + e_i \tag{4.2}$$

Household income, *inc*, is proxied by a dummy for households owning at least one car, another dummy for households owning at least one fridge, and a continuous variable measuring the average amount of money spent by adults when eating outside the household (equal to zero if no household members ever eat outside). Other explanatory variables include household size, *hhsize*, the share of children (below 14 years old) in the household, *kids*, the average number of years of education of adult males, *medu*, and adult females, *fedu*, a dummy variable indicating whether households do (equal to zero) or do not (equal to one) observe fasting practices as indicated by the orthodox Christian calendar, *fast*, and a dummy variable that equals one when female adults are responsible for dairy shopping in the household, *shop*.

Moreover, an interaction term (*kids*fast*) is added to the empirical model in order to capture non-linear effects associated with fasting practices, *fast*. In particular, there might be households that do observe fasting practices, but they still consume dairy during the fasting periods. This happens in households with children below seven years of age who are exempted from fasting practices. To control for this we interacted *kids* with *fast* and we introduced this interaction term in the empirical model.

It is also important to note that the empirical model (Equations 4.1 and 4.2) could suffer from econometric problems inherent to selection bias, and these should be addressed before estimation. Identifiers to control for the selection problem can be found if there are some

---

[28] Equation 4.2 describes the volume of dairy products purchased by a household in supermarkets. Households choose whether to shop for dairy in supermarkets, and thus, whether we observe their dairy purchase in our data. If households made this decision randomly, we could use an ordinary regression to fit a dairy purchase model. Such an assumption of random supermarket choice, however, is unlikely to be true. For example, supermarkets tend to be located in wealthy urban areas and to offer the most expensive dairy products. Households with lower income and located farther from supermarkets are unlikely to choose supermarkets to shop for dairy, so that the sample of observed purchases becomes upward biased, justifying the use of a Heckman model controlling for the selection bias.

variables that strongly influence the chances to shop for dairy in supermarkets, *smp*, but not the volume of dairy purchased, *smq*. In this specific case such identifiers can be dummy variable indicating the sub-cities where households are located. Using three dummies (accounting for differences between the four sub-cities sampled) in Equation 4.1 we can control for spatial fixed effects. The three dummies selected control for the fixed effects of in the richest, *rw*, the middle rich, *mrw*, and the middle poor, *mpw*, *woredas* compared to the poorest *woreda* (which is used as the base). The location dummies also proxy for the distance between households and supermarkets, since supermarkets are largely concentrated in the richest sub-city. In this way we believe we can explain households' choice to shop for dairy in supermarkets or in other outlets, and separate this from the decision related to the quantity of dairy to be purchased in these outlets.

In many cases, when econometric models are built on data collected at one point in time, as in this case, it is difficult to ascertain that right hand side variables cause variations in the left hand side variable rather than the other way around (endogeneity). However, the direction of causality does not seem to be a problem in our analysis as household income, asset, education, demography, organisation, and location are used as explanatory variables. To avoid endogeneity, we excluded variables measuring the price paid by consumers to purchase dairy products. In fact, it is reasonable to argue that the price paid for dairy products depends largely on consumer choice between industrial and traditional dairy products, which is strongly correlated with the choice to shop for dairy in supermarkets rather than alternative outlets (see section three). An additional concern related to the use of cross section data is heteroskedasticity, which in our model was controlled by using robust standard errors.[29]

4.7 Results

The econometric results obtained through the estimation of the Heckman model described in the previous section are displayed in Table 4.5. We discuss these results in two steps, following the structure of the Heckman model used for estimation. We begin by presenting the results related to the probability that a household shops for dairy in supermarkets (last column on the right), reporting also the marginal effects for the significant variables (Table 4.6). First of all, we notice that for consumers located in the richest *woreda* the probability to shop for dairy in supermarkets increases by 58 percent. On the contrary, location dummies for middle-rich and middle poor sub-cities are insignificant, emphasising once more the disproportionate concentration of supermarkets within the richest area of the city, and justifying the use of location dummies to control for selection bias.

Amongst the three income proxies, only owning a fridge is significant in explaining the probability to shop for dairy in supermarkets. Results suggest that owning a fridge increases the probability to shop for dairy in supermarkets by 21 percent. In order to explain this finding

---

[29] Heteroskedasticity occurs when the variance of the random error term is not constant across observations.

Table 4.5. Dairy shopping from supermarkets (Heckman model), Addis Ababa, 2006.

Household characteristics	Volume of dairy products purchased from supermarkets (in milk equivalents)	Probability to shop for dairy in supermarkets
Dummy for owning a car	18.84 (38.54)	0.09 (0.35)
Dummy for owning a fridge	-7.13 (37.91)	0.68 (0.31)**
Money spent for a meal outside (Birr/time)	2.36 (0.71)**	-0.00 (0.01)
Household size	-9.88 (8.90)	0.05 (0.06)
Dependency ratio (no. children/hh size)	192.47 (135.87)	-2.13 (0.82)**
Dummy for nonfasting households	0.84 (35.24)	0.49 (0.27)**
Non fasting households*dependency ratio	-0.26 (0.16)	-0.00 (0.00)
Male adults education (years)	-0.59 (2.99)	0.01 (0.02)
Female adults education (years)	1.95 (3.80)	0.06 (0.03)**
Dummy, female responsible for shopping	1.75 (24.22)	-0.47 (0.16)**
Dummy for location in richest *woreda*	-	1.68 (0.43)**
Dummy for location in middle-rich *woreda*	-	0.56 (0.36)
Dummy for location in middle-poor *woreda*	-	0.12 (0.36)
Inverse Mills ratio	5.20 (0.21)**	
Number of observations	136 censored, 61 uncensored 198	
Log pseudo-likelihood	-460.847	
Wald test of independent equations ($\rho=0$)[1] Chi-square (1) = 6.03 Prob>Chi-square = 0.0140		

Standard error in parentheses (), *denotes significance at 10% level, **denotes significance at 5% level.
[1] The Wald (or likelihood-ratio) test is an equivalent test for the comparison of the joint likelihood of an independent probit model for the selection equation and a regression model on the observed dairy purchasing, as opposed to the Heckman model likelihood. Since Chi-square=6.03, it clearly justifies the use of the Heckman selection model, as opposed to two separate regressions.

it is important to stress some key differences between supermarkets and traditional retailers. Unlike supermarkets, traditional retailers, mainly represented by kiosks, itinerant traders and urban farmers, are small but very numerous and scattered all over the city, often providing home delivery services. These characteristics attract households with no fridge, which need to purchase milk frequently and in small quantities.

With regard to household structure, Table 4.5 highlights the irrelevance of household size as well as the importance of the dependency ratio in determining whether the household

Table 4.6. Marginal effects of significant variables, Addis Ababa, 2006.

Household characteristics	Probability to shop for dairy in supermarkets
Dummy for owning a fridge	0.21 (0.092)
Dependency ratio (no. children/hh size)	-0.62 (0.22)
Dummy for nonfasting households	0.14 (0.08)
Adult female education (years)	0.02 (0.01)
Dummy for adult female in charge of shopping	-0.15 (0.04)
Dummy for location in richest *woreda*	0.58 (0.13)

shops for dairy in supermarkets or not. Since milk is the single most important product for child nutrition, household consumption of dairy is expected to increase with the number of children. As discussed in section three, the milk available in supermarkets is either pasteurised or in powder, and therefore more expensive than the raw milk found outside supermarkets. It follows that to minimise the increase in milk expenditure associated with an additional child, the household may tend to buy cheaper, raw milk from urban farmers. Economic reasons may thus justify the 62 percent reduction in the probability to shop for dairy in supermarkets, for a one unit increase in the dependency ratio.

When women are in charge of shopping, the probability to purchase dairy in supermarkets decreases by 15 percent. Typical Ethiopian households are characterised by a traditional structure, with men employed outside the household and adult females working as housewives. It follows that women have on average more time to allocate to household tasks, and therefore attach less value to the convenience of ready to use dairy products from supermarkets and prefer to purchase cheaper raw milk and fermented butter, to be boiled and filtered before consumption, from traditional retailers.

The results in Table 4.5 further show that the level of education of women is also an important factor explaining the probability to shop for dairy in supermarkets. It is interesting to note that the level of education of men is insignificant, emphasising once more that dairy shopping is mainly a female task, regardless of the education level. One additional year of schooling for women implies two percent increase in the probability to purchase dairy from supermarkets. Educated females are expected to be involved in jobs outside the household (less time for household tasks) and to attach more value to the superior hygiene of dairy products sold from supermarkets. The results also show that the probability to shop for dairy in supermarkets is 14 percent lower in households observing fasting practices prescribed by the orthodox Christian church. Households that observe fasting tend to have a more traditional eating culture, and are therefore keener to maintain the habit to purchase milk and dairy from traders and urban farmer rather than from a fancy supermarket.[30]

The middle column in Table 4.5 shows the results for the quantity of dairy purchased from supermarkets. Only the continuous income variable, the average amount spent by adult household members on a meal outside the household, is found to have a significant effect. When the amount of money paid for a meal increases by one Ethiopian Birr, the volume of dairy purchased from supermarkets increases by 2.4 kg/year. Since all other variables are insignificant in explaining the quantity of dairy consumed from supermarkets, it seems reasonable to conclude that when a consumer chooses supermarkets to shop for dairy, the quantity he or she purchases depends exclusively on the amount of money they allocate on food (i.e. on household food expenditure). On the contrary, other factors than food expenditure power appear to prevail in explaining the probability to shop for dairy in supermarkets, as explained above.

4.8 Conclusions and implications

The story that emerges from this analysis is a story of a quest towards economic development, urbanisation, industrialisation and market integration. Ethiopia, like Kenya and several other developing countries, is witnessing the simultaneous emergence of modern processing industries and supermarkets underpinning radical changes in dairy supply chains. Our econometric results suggest that the trends observed in both manufacturing and retailing sectors will continue as long as the Ethiopian urban economy will keep growing. In particular, the demand for industrial dairy products sold by supermarkets emerges in residential areas where per capita food expenditures are highest, where women are educated and employed outside the household, where households do not observe religious fasting practices, and where households have fridges and only have a few kids.

At the moment, manufacturing industries and supermarkets have only a small share of the Ethiopian dairy market. This is due to widespread poverty and incipient urbanisation, but also to the fact that supermarkets are less easy to access than the numerous smaller shops, and because industrial products are less tasty and nutritious, and have higher prices than traditional (informal) dairy supplies. Unmatched consumer preferences will not be necessarily taken into account by emerging supermarkets and industries, unless competition increases in both retail and industrial sectors. In Addis Ababa, like in many other parts of the world, supermarkets and industries tend to concentrate into a retail-industrial oligopoly, increasing transaction costs, and reducing bargaining power of consumers, and eventually of farmers (since an oligopoly is also an oligopsony at the market for its raw materials).

---

[30] Table 4.5 shows also that the interaction term (fasting household*dependency ratio) is insignificant. However the introduction of this interaction term may be justified by the fact that it improves the significance of the fasting variable.

To address such unmatched consumer preferences and the dangers of farmers' exploitation by the emerging dairy oligopoly, the role of the public sector is to ensure fair competition in the retail and industrial sectors. For instance, governmental agencies and NGOs could encourage the establishment of new supermarkets and dairy processing industries. And they could promote the competitiveness of small retailers and industries by demanding ethical conduct by oligopolists and oligopsonists vis-à-vis consumers and farmers. An important role for scientific research in this regard may be to define institutions that can help farmers and consumers to cope with oligopoly and oligopsony power.

Chapter 5. The impact of collective marketing on milk production and quality: a case study from the dairy belt of Addis Ababa

Abstract

Using primary bio-economic data from Ethiopia, this study evaluates the impacts of a dairy marketing cooperative on milk production, productivity, quality and safety at the farm gate. To do so we compare the performance of cooperative farmers and individual farmers within the same area. We use both instrumental variable regression and propensity score matching approaches to control for observable and unobservable differences between the two groups of farmers. Findings are consistent across the two approaches in suggesting that cooperative membership has a positive impact on milk production and productivity, no significant effect on milk hygiene and a negative impact on milk quality. The study concludes with implications for policy and for further research.

5.1 Introduction

Urbanisation and globalisation processes are inducing turbulent changes in the food markets of developing countries. Besides traditional spot-markets, developing countries are witnessing the emergence of highly integrated value chains led by the rapid rise of supermarkets, as documented in Asia (Chang, 2005; Lee and Reardon, 2005; Hu *et al.*, 2004; Rangkuti, 2003; Thailand Development Research Institute, 2002), Latin America (Balsevich, 2005; Reardon *et al.*, 2005; Berdegue *et al.*, 2005; Orellana and Vasquez, 2004; Hernandez, 2004; Alarcon, 2003; Farina 2002; Gutman, 2002; Reardon and Berdegue, 2002), Africa (Neven *et al.*, 2006; Weatherspoon and Reardon, 2003), as well as in Ethiopia (see Chapter Four).

The emergence of supermarkets and value chains is a great opportunity to boost agricultural growth, rural development and poverty alleviation (Weatherspoon and Reardon, 2003; Reardon *et al.*, 2006). However, as observed also in Ethiopia (see Chapter Four), this opportunity is often counterbalanced by the problems associated with the concentration of market power into industrial/retail oligopolies/oligopsonies, favouring the exploitation of smallholder farmers (see Eagleton, 2006; Reardon *et al.*, 2006; Weatherspoon and Reardon, 2003; Kaplinsky and Morris, 2001).

Risks and opportunities associated with changes in the market enivronment are inducing many developing countries to return to agricultural cooperatives (Cook and Chambers, 2007; World Bank, 2003; Collion and Rondot, 1998). A mainstream argument is that cooperatives can reduce transaction costs and improve the bargaining power of smallholder farmers vis-à-vis increasingly integrated markets (Bonin *et al.*, 1993; Munckner, 1988; Dulfer, 1974). After the downfall of the Derg (socialist) regime, also Ethiopian policy makers have been actively promoting a return to agricultural cooperatives to help smallholder farmers better compete in the marketplace (see also Bernard *et al.*, 2008).

As far as the Ethiopian dairy market is concerned, recent studies (D'Haese *et al.*, 2007; Ahmed *et al.*, 2003; Holloway *et al.*, 2000; Nicholson, 1997) suggest that farmers' participation in marketing cooperatives results in a significant increase in milk production and productivity. The objective of this study is to test the latter hypothesis, as well as to highlight further undocumented impacts of Ethiopian dairy marketing cooperatives. In particular, as different markets provide different incentives for quality, the participation of Ethiopian farmers in dairy marketing cooperatives is expected to induce relevant changes in milk quality attributes at the farm gate, with important implications for consumers, retailers, manufacturers and farmers.

Changes in milk attributes can have relevant implications for consumers' health. Milk is in fact a source of energy, essential amino acids and micronutrients, particularly needed in less-developed countries, where diets are mainly based on staple grains or root crops (Fitzhugh, 1999). Moreover, milk is a perishable product, a potential source of food poisoning and diarrhoeal diseases, as well as other known and unknown human diseases (O'Connor, 1995).[31] Milk hygienic attributes are particularly important in less developed countries where diarrhoeal diseases alone represent the leading cause of illness and death, killing approximately 1.8 million people annually, most of whom are children.[32]

Milk quality and safety attributes have also important economic implications throughout the dairy supply chains (see Weaver and Kim, 2001). In some dairy chains, farmers' milk supplies that do not comply with the standards imposed by downstream firms are rejected, implying potential losses for farmers who may not find alternative markets to sell to, as well as sub-optimal milk provisions for manufacturers and retailers. Even when farmers' milk complies with quality requirements, the higher the quality and safety of milk, the higher the profitability of butter and cheese making, and the longer the shelf life of intermediate and final milk products, with clear advantages for both manufacturers and retailers. In an emerging market, like the one in Ethiopia, quality and safety attributes of supplies are expected

---

[31] Known milk borne infectious diseases are: typhoid fever, scarlet fever, septic sore throat diphtheria, tuberculosis, and brucellosis. Unknown diseases can result from bacteria mutation or cross-contamination. (O'Connor, 1995).

[32] This information was obtained from: http://www.who.int/medacentre/factsheets/fs237/en/

to become increasingly important in driving the decisions of manufacturers and retailers on where to purchase and from whom.[33]

With these motivations we proceed with the assessment of the impact of a major Ethiopian dairy marketing cooperative on both milk production and quality. To do so, section two presents some background information about the production site, the target market, the value chain and the dairy marketing cooperative. Section three describes the sample and data available. Section four defines the two analytical methods (regression analysis and propensity score matching) used to control for observable and unobservable differences between cooperative farmers and neighbouring individual farmers. The last two sections discuss the findings using references from previous relevant literature, and derive implications for policy and further research.

5.2 Background

According to the Ethiopian proclamation number 85 from 1994, cooperatives are defined as 'associations established by individuals on a voluntary basis, to collectively solve economic and social problems and to democratically manage them'. In order to achieve legal recognition a cooperative cannot have less than ten members, who must have no major irregularities in their financial records. While the distribution of property rights among cooperative members has been deregulated by a subsequent proclamation (number 147 from 1998), decision making processes in cooperatives remain tied by law to the principle of one member one vote.

The cooperative selected for this study is located in the milk-shed of Debre Zeit, 50 km south of the capital Addis Ababa. In addition to the cooperative and its 800 members (the second biggest dairy cooperative of Ethiopia), this area comprises more than 1000 small dairy farmers (according to the Ministry of Agriculture), a few large dairy farms, two dairy processing plants, and the experimental dairy unit of the International Livestock Research Institute (ILRI). Overall the milk-shed of Debre Zeit represents the most important production-site of Ethiopia, key source of dairy for the market of Addis Ababa.

The milk-shed of Debre Zeit is located on the border between the central Ethiopian plateau and the Rift Valley, at an altitude of approximately 1600 meters above sea level. Biophysical attributes, like the availability of vast grazing areas, mild slopes and environmental temperatures (0-30 °C), and adequate rainfalls patterns (1000-1900 mm/year) offer a relatively disease-free environment with high potential for animal feeding and for the introduction of high-yielding dairy cows (Ahmed *et al.*, 2003). Besides production potential, the milk shed of Debre Zeit is also witnessing increasing demand pressure from the market of Addis Ababa, where industries and supermarkets are rapidly growing (see Chapter Four).

---

[33] Ethiopia is one of the eight additional least-developed countries (LDCs) in the process of accession to the WTO (http://www.wto.org/english/thewto_e/whatis_e/tif_e/org7_e.htm).

According to primary and secondary (Tegegne, 2003) information, the *Ada'a Liben Woreda Dairy and Dairy Products Marketing Association* was established in Debre Zeit in 1997-98 by 34 retired military officers of the national Air Force (also located in Debre Zeit). The current manager of the cooperative is an ex colonel of the Air Force with a master degree in business administration, and the cooperative board is still dominated by military officers from the original group of founders. In the last nine years the number of cooperative members has increased considerably to almost 800 (Figure 5.1). Such a rapid growth in the number of members depends on pulling factors, among which subsidised services, facilitated credit and donations play a dominant role.

The policy of the cooperative states that any individual has the right to join in, as long as he/she can afford to pay the entrance fee, and to purchase at least one share of the collective endowment. Fees and shares are set on the basis of regular internal evaluations, are redeemable but cannot be traded, not even among members. Furthermore, a fixed percentage (10 percent) of members' revenue (generated by selling milk through the coop) is retained as a form of patronage to build up additional equity capital and cover running costs.

The cooperative has a federated structure based on 11 collection centres placed in strategic production sites in and around the town of Debre Zeit. On a daily basis, cooperative farmers deliver whole raw milk to these centres, share information, and procure inputs. Like most cooperatives in the country, this cooperative represents a preferential nexus between dairy farmers and public subsidies to procure artificial insemination and exotic or cross-bred cows.

Output services involve quality control, milk collection and bulking, cooling and processing, transportation and commercialisation (all activities are undertaken twice a day, seven days a week). Before collection, all milk supplies (only from coop-members) are screened using instantaneous tests (alcohol test and specific gravity test), which measure milk quality as good or bad, but cannot provide a continuous scale of grades.[34] Milk supplies that do not comply with the minimum standards set by these tests are rejected, even if the rejection rate appears to be negligible. Approved milk supplies are weighted, recorded and bulked.

Approximately 50 percent of the milk collected by the cooperative is sold from the collection centres directly to local consumers, or transformed into cottage cheese (*ayb*), sour milk (*ergo*) and fermented butter (*kebe'*), without undergoing any other heat treatment or packaging process. The other half of the milk supply is immediately transported (without cooling facilities, and at the expense of the cooperative) and transferred to manufacturing firms upon

---

[34] *The alcohol test* is a low cost, instantaneous technique to evaluate the status of the milk colloidal suspension. When one part of alcohol is added to one part of milk with major alterations in the colloidal suspension, the solution precipitates, indicating that the milk is old or contaminated (O'Connor, 1995). *The specific gravity test* is a low cost, instantaneous technique to compute milk density, given milk temperature (O'Connor, 1995). This test makes use of a floating device and a thermometer, and allows to infer about major variations in fat and protein content, in particular those associated to water addition and/or cream removal.

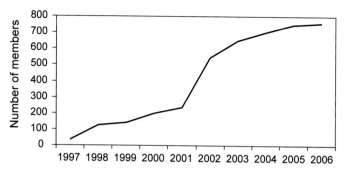

Figure 5.1. Growth in cooperative membership over time, Debre Zeit.

compliance with quality standards and other written or verbal agreements. These firms produce pasteurised and packed products such as whole, partially skimmed, and skimmed milk, butter, cottage and cheddar cheese, yoghurt, etc., which are then distributed to supermarkets, and to a lesser extent to kiosks and specialised dairy shops, in Addis Ababa.

5.3 Data

Within the *woreda* (i.e. municipality) of Debre Zeit, we surveyed 50 cooperative farmers and 50 individual farmers. Each farmer was interviewed using a structured questionnaire to collect information on farm-household characteristics. Two samples were also collected from the milk bulk of each farmer interviewed, and subsequently analysed using standard chemical and microbiological tests. The available dataset provides an unique combination of biological, technological and socio-economic information about dairy farming in Ethiopia.

The sampling method can be described as follows. In July-August 2003, 20 cooperative farmers and 20 individual farmers were randomly selected, on the basis of a list provided by the Ministry of Agriculture. The sampling area included the urban area of Debre Zeit and the peasant association (i.e. rural community) of Babogaya. Each site included both coop-members and individual farmers, even if in different proportions.

In order to expand the number of observations we conducted a second survey round in 2006. Due to difficulties in tracing the same farmers interviewed in 2003 (some quit the cooperative, others died or had dry cows, etc.), cooperative farmers (30) and individual farmers (30) were again randomly selected from the list and the sampling areas considered in 2003. As in 2003, the second survey was conducted between July and August in order to minimise seasonality effects. It is also important to note that this second survey was conducted using the same enumerators, identical questionnaires and milk sampling procedures as in the first survey.[35] Finally, all milk samples collected in 2003 and 2006 were analysed in the laboratory of ILRI Debre Zeit, by the same technicians, using identical grades and standards.[36]

5.4 Methodology

The first step in impact analysis is to select appropriate impact indicators. In this study, milk production is defined as the average quantity (litres) of milk produced by a farm, on a daily basis, during the last year. Milk productivity is defined as the ratio between milk production and the number of milking cows available per farm. Given available data, milk quality and safety are here stylised using standard laboratory grades for fat and protein content, and total bacterial count. The selection of these indices is justified as follows. Milk is a complex emulsion with high density of nutrients (Walstra, 2006). Variability among milk components is largely inter-dependent, but the widest variations occur in fat and protein content (gr/ml), making of these the most common indices for the quantitative evaluation of milk nutritional value (O'Connor, 1995).[37] Moreover, milk is an ideal terrain for bacterial growth, a condition that makes the total bacterial count (TBC, measured in cfu/ml) a widely used test for the quantitative evaluation of general milk hygiene (O'Connor, 1995).[38]

Given these performance indicators, the objective of this study is to compute the average treatment effect on the treated (ATT), i.e. the impact of cooperative membership on the performance of cooperative farmers. As posed by Ravallion (2001) and Godtland et al. (2004) the empirical problem we face is the typical absence of data concerning the counter-factual: how would the performance of cooperative farmers have been if these farmers had not joined the cooperative? Our challenge is to identify a suitable comparison group of non-cooperative farmers whose performance – on average – provides an unbiased estimate of the performance that cooperative members would have had in the absence of the cooperative. However, due to farmer self-selection and the sample design (cooperative and individual farmers were sampled from the same area), there are three potential sources of bias in comparing performances of cooperative and individual farmers.

The first source of bias is related to diffusion or spill-over effects across cooperative and individual farmers. In particular, since the two groups of farmers were selected from the same

---

[35] In both surveys (2003-2006), milk samples were gathered and analysed within a one-month period, so as to reduce the influence of climate variations. Sampling steps: sanitise the equipment (planger and diper) with running water, and operator hands with alcohol (70 percent); stir milk bulk; collect a milk sample and pour it into a sterile container properly labeled; immediately store the sample in an icebox (0-4 °C).

[36] Total bacterial count (Standard Plate Agar), Milk Fat Content (Gerber method), and Milk Protein Content (Protein Formaldehyde Titration).

[37] Interdependent variability means that modifications in one component affect most of the other components.

[38] cfu = colony forming units. The higher the number of cfu per ml of milk, the faster the spoilage process in milk and related products.

woreda (i.e. municipality), the comparison of these two groups is likely to underestimate cooperative's impacts. As discussed above, in Ethiopia and especially in this specific case, cooperative membership means access to subsidised agricultural inputs as well as to emerging market opportunities. Inputs and buyers attracted by the cooperative could somehow trickle down to neighbouring individual farmers. For instance, unlike in other parts of the country, in the area surrounding this cooperative we observe the presence of private services, outside the cooperative framework, providing improved bulls for natural insemination, veterinary care, and feed supplies. On the output side, although the cooperative does not purchase milk supplies from non-members, industrial plants and collection centres of private manufacturing firms are growing in the area, offering additional outlets also to individual farmers. Since diffusion bias cannot be controlled for or even estimated, with the data available, we can only acknowledge its likely presence and therefore the possibility that our findings underestimate cooperative's impacts.

The second and third sources of bias are related to selection on observables and unobservables. Given farmer self-selection in the cooperative, a simple comparison of performance indicators between participants and non-participants (naïve analysis) would yield biased estimates of cooperative impact. Coop-members are likely to differ from individual farmers in the distribution of observable (such as age, education, household composition, etc.) and unobservable characteristics (e.g. farmer's ability and motivations that explain both the decision to join the cooperative and farming performance). These differences must be taken into account in comparing the two groups, since they might have had an influence on performance even in the absence of the cooperative.

Following Godtland *et al.* (2004) and Bernard *et al.* (2008), the observable characteristics included in this analytical model are: the level of formal education (in years) and the age (in years) of the household member responsible for dairy production; household size; and the percentage of children (below 14 years old) and women in the household. Figures 5.2a-e show that even if there are differences in the distribution of observable characteristics across cooperative and individual farmers, a common support (area where box-plots overlap) is observed between the two groups, for each and every characteristic. This provides the basic conditions to justify the use of individual farmers as control group.

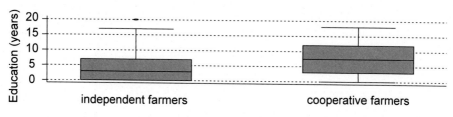

Figure 5.2a. Education of the household member responsible for dairy, Debre Zeit, 2003/06.

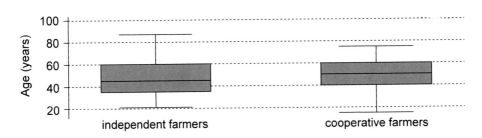

Figure 5.2b. Age of household member responsible for dairy, Debre Zeit, 2003/06.

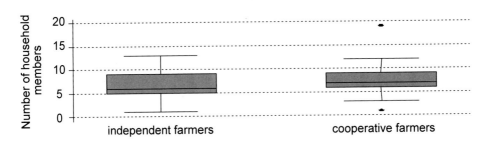

Figure 5.2c. Household size (number of members), Debre Zeit, 2003/06.

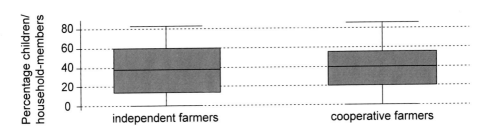

Figure 5.2d. Percentage of children < 14 years old in the household, Debre Zeit, 2003/06.

Figure 5.2e. Percentage of women in the household, Debre Zeit, 2003/06.

In order to control for selection bias due to unobservable characteristics we need instrumental variables that explain the decision to participate in the cooperative, but do not influence the performance given participation. Given this basic principle and available data we identified two valuable instruments. The first is a dummy for participation in the military Air Force. This instrument is theoretically justified since the cooperative was originally established by a group of retired military officers, previously employed by the national Air Force based in Debre Zeit, who are in charge of cooperative management. Table 5.1 shows that 42 percent of cooperative households comprise an officer (or ex-officer) of the Air Force, against the 10 percent of individual farm-households. This may imply that households affiliated with the Air Force have better access to information about the cooperative, as well as better incentives to join in.

The second instrument is a dummy for rural (equal to one) or urban (equal to zero) location of farm-households. Table 5.1 shows that 51 percent of individual farms are located in rural areas, as opposed to only eight percent of cooperative farms. Since cooperative centres for milk collection and input distribution are located within or in proximity of the urban area, rural farm households have less incentives to join the cooperative. Tables 5.2, 5.3 and 5.4 suggest that the choice of instruments is also statistically justified. While military affiliation has a positive and significant (10 percent level) impact on cooperative membership (Table 5.4), the correlation between military affiliation and performance indicators is not significant, given participation (Tables 5.2 and 5.3). Similarly, rural location has a negative and significant (5 percent level) effect on membership (Table 5.4), but no clear correlation with any performance indicators (Tables 5.2 and 5.3).

Given these important considerations we proceed with the estimation of cooperative's impact using two separate methods: instrumental variable regression (IVREG) and propensity scores matching (PSM). For both methods, the Probit model presented in Table 5.4 is a necessary first step. In order to support the robustness of IVREG analyses we excluded a few influential observations (outliers) from each regression; we made sure that all residuals estimated were normally distributed, and we applied robust standard errors to control for heteroskedasticity

Table 5.1. Instrumental variables, Debre Zeit, 2003/06.

Groups\variables	Military affiliation				Rural location			
	Mean	Stand. dev.	Min.	Max.	Mean	Stand. dev.	Min.	Max.
Cooperative farmers [50 obs.]	0.42	0.50	0	1	0.08	0.27	0	1
Individual farmers [50 obs.]	0.10	0.31	0	1	0.51	0.51	0	1

Table 5.2. Correlation between productive performance and instruments, Debre Zeit, 2003/06.

	Production	Productivity
Cooperative membership	9.56 (1.02)**	5.09 (0.84)**
Dummy for survey year	0.86 (0.97)	0.87 (0.59)
Observable characteristics		
Farmer education	0.32 (0.25)	-0.12 (0.15)
{Farmer education}^{2}	-0.02 (0.01)	0.01 (0.01)
Farmer age	-0.10 (0.14)	-0.05 (0.13)
{Farmer age}^{2}	0.00 (0.00)	0.00 (0.00)
Household size	0.44 (0.55)	0.05 (0.27)
{Household size}^{2}	-0.04 (0.04)	-0.01 (0.01)
% of children	1.32 (2.11)	0.67 (1.13)
% of women	0.48 (3.08)	-1.06 (1.88)
Unobservable characteristics		
Dummy for military affiliation	-1.53 (1.18)	-0.30 (0.73)
Dummy for rural location	-1.64 (0.99)	-0.41 (0.77)
R-squared	0.6618	0.5309
N. of observations	89	92

Standard error in parentheses (), *denotes significance at 10% level, **denotes significance at 5% level.

(detected in all regressions). On the other hand, in order to improve the robustness of the PSM method, we restricted matches to farmers with propensity scores within the area of common support (as defined by Smith and Todd, 2000). Consequently, block identifiers are missing for control observations outside the common support and the number of valid observations reduces from 100 to 91 (see Table 5.5).[39] Statistical robustness of PSM is further supported by the use of two different matching techniques (Kernel and Stratification method).[40] Finally, it is important to clarify that both IVREG and PSM methods provide unbiased measures of

---

[39] Note that STATA reports that balancing property is satisfied across the block identifiers of the propensity scores estimated.

[40] Nearest Neighbour and Radius matching methods were not used since they would have discarded observations from an already small sample (Becker and Ichino, 2001).

Cooperation for competition

Table 5.3. Correlation between qualitive performance and instruments, Debre Zeit, 2003/06.

	Dependent variables		
	Fat content	Protein content	TBC
Cooperative membership	-1.59 (0.31)**	-0.47 (0.10)**	-1.34e+07 (7.06e+06)*
Dummy for survey year	-0.32 (0.24)	-0.12 (0.09)	-6.45e+07 (7.29e+06)**
Observed characteristics			
Farmer education	-0.01 (0.06)	-0.00 (0.02)	1.61e+06 (2.41e+06)
{Farmer education}^{2}	0.00 (0.00)	0.00 (0.00)	-1.95e+06 (1.39e+05)
Farmer age	0.06 (0.04)	0.02 (0.01)	-1.33e+06 (1.66e+06)
{Farmer age}^{2}	-0.00 (0.00)	-0.00 (0.00)	1.15e+04 (1.57e+04)
Household size	0.02 (0.08)	-0.01 (0.03)	3.21e+06 (3.51e+06)
{Household size}^{2}	0.00 (0.00)	0.00 (0.00)	-1.56e+05 (1.54e+05)
% of children	-1.17 (0.65)*	-0.22 (0.22)	-5.62e+06 (1.71e+07)
% of women	-2.27 (0.72)**	-0.18 (0.35)	1.35e+07 (2.63e+07)
Unobserved characteristics			
Dummy for military affiliation	0.17 (0.27)	0.02 (0.09)	2.07e+06 (7.20e+06)
Dummy for rural location	0.29 (0.30)	0.04 (0.11)	-9.44e+06 (7.82e+06)
R-squared	0.4261	0.2866	0.5583
N. of observations	94	97	98

Standard error in parentheses (), *denotes significance at 10% level, **denotes significance at 5% level.

cooperative impacts under the assumption of non-simultaneity (i.e. non-endogeneity) across performance indicators and explanatory variables.[41]

---

[41] Note that in order to improve predictions, the probit regression presented in Table 5.4 (and consequently also correlations in Tables 5.2 and 5.3, and instrumental regressions in Tables 5.6 and 5.7) are intentionally over-parametrised, using as many variables and quadratic terms as possible. Note also that besides observable and unobservable characteristics, all regressions and correlations include a dummy indicating the year in which households were surveyed (equal to zero for 2003 and one for 2006). This dummy allows to control for eventual inconsistencies between the two surveys.

Table 5.4. Probability of cooperative membership (Probit), Debre Zeit, 2003/06.

Explanatory variables	Cooperative membership	
	Probit	Marginal effects
Dummy for survey year (2006=1 & 2003=0)	-0.10 (0.30)	-0.04 (0.12)
Observable household characteristics		
Farmer education (years)	0.14 (0.09)	0.05 (0.04)
{Farmer education}^{2}	-0.00 (0.01)	-0.00 (0.00)
Farmer age	0.13 (0.07)**	0.05 (0.03)**
{Farmer age}^{2}	-0.00 (0.00)*	-0.00 (0.00)*
Household size	0.14 (0.15)	0.06 (0.06)
{Household size}^{2}	-0.01 (0.01)	-0.00 (0.00)
% of children (< 14 years old)	-1.25 (0.76)*	-0.50 (0.30)*
% of women	-0.98 (1.00)	-0.39 (0.40)
Unobservable household characteristics		
Dummy for military affiliation	0.63 (0.39)*	0.25 (0.14)*
Dummy for rural location	-1.21 (0.40)**	-0.44 (0.12)**
Pseudo R-squared	0.3132	
Log pseudolikelihood	-46.65	
Correctly classified observations	77.55%	
N. of observations	98	

Standard error in parentheses (), *denotes significance at 10% level, **denotes significance at 5% level.

Table 5.5. Blocks of propensity scores, Debre Zeit, 2003/06.

Propensity scores' blocks	Individual	Cooperative	Total
0.045	15	1	16
0.2	7	5	12
0.4	11	12	23
0.6	7	10	17
0.8	2	21	23
Total	42	49	91

5.5 Results

The discussion of empirical findings begins with the Probit estimation presented in Table 5.4 (previous section). This table shows that besides military affiliation and rural location, also farmer age and household dependency ratio are significant in explaining membership. In particular the probability of being a member of the cooperative increases with farmer age up to a certain threshold, after which the relationship turns negative. Further, cooperative membership becomes less likely when the percentage of children in the household increases. Overall, these findings suggest that personal motivation, based on the physical distance to cooperative infrastructures and social network (or social capital), farmer age and household dependency ratio (interpretable as proxies for labour availability and skill), are key factors in promoting farmers' collective action within an otherwise homogeneous population.

On the basis of these results we predict the probability of cooperative membership, critical to the IVREG analyses presented in Tables 5.6 and 5.7, and estimate propensity scores for matching cooperative with otherwise similar individual farmers (PSM method). Cooperative

Table 5.6. The production impact of cooperative membership (IVREG), Debre Zeit, 2003/06.

Explanatory variables	Production	Productivity
Dummy for survey year	0.81 (1.00)	0.90 (0.57)
Instrumented variable		
Cooperative membership	11.50 (1.90)**	5.69 (1.21)**
Observable characteristics		
Farmer education	0.19 (0.27)	-0.17 (0.17)
{Farmer education}^{2}	-0.01 (0.01)	0.01 (0.01)
Farmer age	-0.13 (0.16)	-0.07 (0.15)
{Farmer age}^{2}	0.00 (0.00)	0.00 (0.00)
Household size	0.35 (0.64)	-0.01 (0.26)
{Household size}^{2}	-0.04 (0.04)	-0.01 (0.01)
% of children	1.84 (2.37)	0.83 (1.22)
% of women	0.49 (3.29)	-0.96 (1.86)
Instruments		
Dummy for military affiliation	/	/
Dummy for rural location	/	/
R-squared	0.636	0.5248
N. of observations	89	92

Standard error in parentheses (), *denotes significance at 10% level, **denotes significance at 5% level.

Table 5.7. The quality impact of cooperative membership (IVREG), Debre Zeit, 2003/06.

Explanatory variables	Fat content	Protein content	TBC
Dummy for survey year	-0.32 (0.25)	-0.13 (0.09)	-6.35e+07 (7.67e+06)**
Instrumented variable			
Cooperative membership	-2.00 (0.66)**	-0.54 (0.24)**	8.95e+06 (1.72e+07)
Observable characteristics			
Farmer education	-0.01 (0.07)	0.00 (0.00)	7.78e+06 (2.65e+06)
{Farmer education}^{2}	0.00 (0.00)	0.00 (0.00)	7.04e+03 (1.41e+05)
Farmer age	0.08 (0.05)*	0.02 (0.02)	-2.08e+06 (1.67e+06)
{Farmer age}^{2}	-0.00 (0.00)	-0.00 (0.00)	1.83e+04 (1.56e+04)
Household size	0.03 (0.09)	-0.01 (0.00)	2.39e+06 (3.41e+06)
{Household size}^{2}	0.00 (0.00)	0.00 (0.00)	-1.28e+05 (1.45e+05)
% of children	-1.28 (0.67)*	-0.23 (0.23)	7.44e+05 (1.94e+07)
% of women	-2.35 (0.75)*	-0.20 (0.34)	1.82e+07 (2.53e+07)
Instruments			
Dummy for military affiliation	/	/	/
Dummy for rural location	/	/	/
R-squared	0.4031	0.3566	0.5109
N. of observations	94	97	98

Standard error in parentheses (), *denotes significance at 10% level, **denotes significance at 5% level.

impacts estimated with IVREG regressions and PSM approach (further distinguished into Kernel and Stratification techniques) are then summarised in Table 5.8 together with the description and naïve comparison (based on t-tests) of production and quality performance of cooperative and individual farmers.

Turning to the results, Table 5.8 shows that the average cooperative farmer produces almost 17 litres of milk per day, with a productivity of eight litres per cow per day. Cooperative milk is characterised by an average 3.6 percent of fat content, 3.0 percent of protein content, and 25 million cfu/ml. On the other hand, the average individual farmer produce 3.5 litres, with a productivity of 2.5 litres, 5.2 percent of fat content, 3.5 percent of protein content, and 31 million cfu/ml.

Table 5.8 shows also that naïve estimates (based on t-tests) do not differ much from the results obtained with PSM and IVREG methods, suggesting that cooperative membership is almost randomly distributed within the sample. Nonetheless, if we exclude the IVREG estimate for

Table 5.8. The impact of cooperative membership (PSM+IVREG), Debre Zeit, 2003/06.[1]

Performance	Cooperative farmers	Individual farmers	Naive (t-test)	ATT kernel	ATT stratification	IVREG
Production (lt/farm/day)	16.8 (11.1)	3.5 (3.3)	13.3[1.7]**	13.7[1.8]**	13.6[1.7]**	11.5[1.9]**
Productivity (lt/cow)	8.0 (6.1)	2.5 (2.5)	5.5[1.0]**	5.8[1.0]**	5.8[1.0]**	5.7[1.2]**
Fat (%)	3.6 (0.6)	5.2 (1.8)	-1.5[0.3]**	-2.0[0.7]**	-1.8[0.6]**	-2.0[0.7]**
Protein (%)	3.0 (0.3)	3.5 (0.6)	-0.5[0.1]**	-0.5[0.2]**	-0.6[0.2]**	-0.5[0.2]**
TBC (cfu/ml)	2.5e+07 (4e+07)	3.1e+07 (4.4e+07)	-5.4e+06[8.4e+06]	8e+05[8e+06]	2.1e+06[1e+07]	8.9e+06[1.7e+07]

Standard deviation in (), Standard error in [], *denotes significance at 10% level, **denotes significance at 5% level.
[1] ATT is equal to the outcome of cooperative farmers minus the outcome of individual farmers after Propensity Score Matching. Since analytical standard errors are not computable for the Kernel and Stratification methods, we compute robust standard errors using 100 bootstrap replications.

milk production, naïve estimates appear slightly but consistently smaller than the estimates computed with IVREG and PSM methods. Consistent underestimation by naïve analyses suggests that, even if small, selection bias might indeed be present in the sample. Considering also that none of the estimation methods allows to control for diffusion (or spill-over) bias (a potential source of additional underestimation), the most realistic impact estimates are expected to be the largest ones (in bold in Table 5.8).

Regardless of the estimation method used, Table 5.8 clearly suggests that the cooperative has a positive impact on milk production and productivity, a negative impact on milk nutritional value (fat and protein content) and an insignificant impact on milk hygiene (total bacterial count). It is important to note that cooperative impact on milk hygiene could have been overshadowed by inconsistencies in milk sampling procedures, or even by a significant change in environmental hygiene, between 2003 and 2006. In support to this hypotheses, Table 5.7 shows that only the dummy indicating the survey year is relevant in explaining milk hygiene (while it is insignificant in explaining the dependent variables in both Table 5.6 and 5.7). Hence, the potential presence of bias due to external factors combined with the small number of available observations induce us to avoid commenting on potential reasons behind the lack of impact of cooperative membership on milk hygiene. Milk hygiene apart, the impacts estimated confirm the two hypotheses presented in the introduction, proving that indeed the formation of dairy marketing cooperatives can induce the improvement of milk production and productivity, as well as modifications in milk quality.

Likely explanations for these findings can be found in the different incentives faced by cooperative and individual farmers. As discussed above, a key difference is associated with the fact that this cooperative, like most Ethiopian dairy cooperatives, provides smallholder farmers with access to subsidised inputs. Subsidies involve mainly the procurement of artificial insemination services and live cows. As a result, cooperative herds are dominated by high yielding crossbred cows, as opposed to the zebu cattle typically found in the herds of non-cooperative farmers. While indigenous zebu cattle are characterised by the production of small volumes of milk (2-3 lt/day) with high density of nutrients, crossbred cows produce larger volumes with lower fat and protein content (Taneja and Aiumlamai, 1999; Walstra *et al.*, 2006). Hence, a great deal of the cooperative impact can be referred to technical innovation through the adoption of crossbred cows. Such an explanation finds large support in the development literature where institutional reorganisation (such as collective action) is typically described as a necessary step towards technological innovation (see Dulfer, 1974; Hayami and Otsuka, 1992; Munckner, 1998).

However, innovation in herds' genotype does not provide an exhaustive explanation for the drastic reduction in nutrient density observed in cooperative's milk. According to dairy literature, the average nutritional value and hygiene estimated in this study for cooperative milk fall largely below most common international standards for milk hygiene, fat and protein content.[42] Hence, the cooperative impact on milk quality must involve additional explanations.

In particular, we observe that cooperative farmers feed their cross-bred cows mainly with dried forages and crop residues, suggesting that the lack of more nutritious, concentrated feed could be a reason for the poor nutritional value in milk yields.[43] Cooperative farmers argue in fact that concentrated feed is scarce and far too expensive, and cooperative managers are constantly in search of affordable feed to redistribute to their members. Moreover cross-bred cows are kept almost constantly inside the barn, suggesting that the lack of movement and grazing could also be part of the problem. On-barn husbandry is a consequence of the fact that most cooperative farms are located within the urban area, as well as by the farmers' fear that something may happen to their precious cross-bred cows while grazing out of sight. Another explanation could also involve poor farm hygiene and inappropriate milking practices favouring the occurrence of mastitis among cross-bred cows, which compared to indigenous zebu cattle are far less resistant to infections. To improve farm hygiene and overall husbandry skills the cooperative is providing some training to its members, but so far only a few members were actually trained.

Last but not least, the poor nutritional value of milk produced by cooperative farmers could reflect inadequate incentives for quality at the farm gate. In particular, we observe that the cooperative makes use of alcohol and specific gravity tests to screen milk quality and safety at the farm gate. The widespread perception among cooperative managers, extension agents and policy makers is that these tests are simple, cheap but also helpful in promoting on-farm quality management. However, modern management theory (Weaver and Kim, 2001) demonstrates that quality control techniques of the type adopted by these cooperatives are often useless.

In support to management theory we observe that specific gravity tests are usually conducted without accounting for differences in the temperature of milk supplies, which is critical to provide a reliable estimation of milk density. Hence the density measures provided by these tests are usually influenced by whether cows were milked five minutes or two hours before. Second, quality control by Ethiopian dairy cooperatives is typically conducted in the absence of independent arbitrage, i.e. in the absence of a third party (public or private institutions, or even anti-trust cooperative bodies) recognised by farmers and cooperative management to certify the legitimacy of milk evaluation processes. This shortfall allows for fraud against buyers and consumers (meaning that milk supplies that do not comply with quality standards are nonetheless accepted and commercialised by the cooperative), or even against farmers (good supplies are rejected), especially against those farmers that are disliked by the rest of the

---

[42] Taneja and Aiumlamai (1999) and Walstra et al. (2006) report average values of 4.8 fat content and 3.2 protein content for zebu cattle and 3.9 fat content and 3.5 protein content for Frisian cows. Walstra et al. (2006) reports average values of 2 million cfu/ml of raw milk.

[43] These arguments are supported by standard dairy literature (see Balasini, 2000; and Belavadi and Niyogi, 1999).

cooperative or simply by the technician that carries out the quality control. Third, alcohol and specific gravity tests are not attribute specific, in the sense that they do not provide cooperative members with precise information about the cause of eventual quality alteration. Finally, these two tests measure milk quality as good or bad, and not on a continuum, hindering the possibility to upgrade quality standards as milk quality improves, nor to set progressive premium prices to promote progressive quality improvement. The uselessness of these tests, is further confirmed by the negligible share of milk supplies that are rejected by the cooperative, suggesting that the quality standards set by these tests lie below the actual (very poor) quality and safety of milk supplies.

5.6 Conclusions and implications

This study evaluates the impact of a prominent dairy marketing cooperative on the performance of smallholder farmers in Ethiopia. To do so it compares a group of cooperative farmers and a group of otherwise similar farmers on the basis of their milk production and quality. Empirical findings suggest that cooperative membership has a positive impact on milk production and productivity, a negative effect on milk quality and an insignificant impact on milk hygiene. The robustness of these findings is demonstrated by the fact that two separate approaches, based on instrumental variable regressions and propensity scores matching, yield consistent results. Furthermore, estimated impacts on production and productivity appear in line with consolidated theory from development economics (Bonin *et al.*, 1993; Dulfer, 1974; Munckner, 1998), and with empirical evidence from previous studies on dairy marketing cooperatives in Ethiopia (Holloway *et al.*, 2000; Ahmed *et al.*, 2003; D'Haese *et al.*, 2007). On the other hand, the effects of cooperative membership on milk quality are far less documented in literature, yet we were able to show that they are also relevant for the profitability and competitiveness of farmers, manufactuers and retailers, and for consumer welfare.

To a large extent, we attribute productivity gains and quality losses to the fact that dairy marketing cooperatives provide access to subsidies for artificial insemination and live animals in Ethiopia, ultimately promoting farmers' adoption of cross-bred cows. As smallholders become cooperative members they shift from indigenous zebu cattle to cross-bred cows, so that milk yields expand in volume but also become more diluted. As extensively documented in dairy science (Walstra *et al.*, 2006), productivity gains, obtained through breeding techniques, are typically associated with significant reduction in nutrient density, especially when feeding and husbandry techniques do not develop in accordance with animal genotype. In line with mainstream dairy science, we observe that cooperative services (for procurement and distribution, information, training, etc.) in support of animal feeding and husbandry are at an infant stage, and also that cooperative farmers do not receive any incentive to improve feeding and husbandry techniques. Consequently, we observe that cooperative milk is characterised by excessive bacteria contamination and insufficient fat and protein content, compared to most common international milk standards (Taneja and Aiumlamai, 1999; Walstra *et al.*, 2006).

In brief, this study indicates that marketing cooperatives can be an institutional panacea in order to promote agricultural growth in Ethiopia. This finding supports national policy-makers, who have been strongly promoting the organisation of smallholder farmers into market-oriented cooperatives since 1994. However, this study also shows that like most cures, marketing cooperatives can cause side effects, which deserve to be carefully examined and addressed.

Chapter 6. Incentives for quality in dairy cooperatives: evidence and implications from the Ethiopian Highlands

Abstract

Using bio-economic data from Ethiopia, we present operational recommendations to optimise milk quality and safety in dairy cooperatives, so that Ethiopian smallholder farmers can better compete in the marketplace. This study shows that nutritional value and hygiene are extremely poor in the milk produced by Ethiopian cooperative farmers. Econometric results suggest that for a given market environment and production technology, milk quality and safety can be still improved through the reorganisation of the structure, services, and grades and standards adopted by cooperatives. The paper concludes with implications for policy and for further research.

6.1 Introduction

Population growth, urbanisation and income growth in the Middle East, North and sub-Saharan Africa are occasioning a massive increase in demand for food of animal origin (Delgado *et al.*, 1999). In particular, FAO-IFPRI-ILRI projections indicate that dairy consumption is estimated to grow by an average 3.8 percent per year in the sub-Saharan region, and by three percent in North Africa and the Middle East.[44] Although making dairy available and accessible remains the most pressing challenge in these regions, research and development efforts cannot neglect the importance of milk quality and hygiene.

Public and private quality specifications are becoming more stringent in the global food market (World Bank, 2007: 156; Mainville *et al.*, 2005; Reardon and Barrett, 2000). However, food quality is still widely perceived as an unaffordable luxury in many developing countries, and as such is typically ignored in development literature. To many, efforts to improve food quality in developing countries are a waste of resources that could be used to alleviate food insecurity, or even a constraint to food production intensification, and therefore to rural poverty reduction. We argue instead that quality is the lubricant of commercialisation, and therefore a necessary element to intensify production. We argue also that no food security can be achieved if diets are based on unsafe products with poor nutritional value.

---

[44] It is important to note that the projected growth rate for sub-Saharan Africa is the second largest in the world after India (4.1%) (Delgado *et al.*, 1999).

In less-developed countries, where diets are mainly based on staple grains or root crops, milk is a key source of calories, essential amino acids and micronutrients (Fitzhugh, 1999). However, milk's precious content varies depending on animal genotype and farming practices, and can be easily adulterated, by adding water or removing cream. Milk is also an ideal terrain for bacterial growth and a potential source of food poisoning and diarrhoeal diseases, as well as other known and unknown infections (O'Connor, 1995).[45] Milk hygiene is particularly important in less developed countries where diarrhoeal diseases alone represent the leading cause of illness and death, killing approximately 1.8 million people annually, most of whom are children.[46] For this set of reasons, Europe and the US (see EUFIC[47] and FDA[48]) have enforced quality and safety standards to regulate national milk trade, and developing countries are expected to follow, encouraged by increasing international competition and health concerns.

Milk quality has important implications also for production and poverty. Milk is typically marketed through value-adding, cold chains, where the profit and competitiveness of downstream chain operators depends on the quality performance of upstream farmers (see Weaver and Kim, 2001). Intuitively, the higher the quality and hygiene of milk at the farm gate, the higher the profitability of butter and cheese making, and the longer the shelf life of intermediate and final dairy products, with clear benefits for downstream manufacturers, retailers and consumers.

As a result, farmer participation in dairy supply chains is increasingly regulated with private grades and standards. In some cases raw milk supplies that do not comply with the quality standards imposed by downstream buyers are rejected from the supply chain. Consequently, farmers may not find alternative markets to sell to, or end up selling to casual buyers outside the supply chain. In other cases no milk is excluded from the supply chain, but milk price is set accordingly to quality grades, calling for increasingly efficient practices to improve milk quality.

In several developing countries milk is produced by smallholder farmers organised in cooperatives (World Bank, 2007: 154; Ahmed *et al.,* 2003; Staal, 1995). Agricultural cooperatives can improve access to input and output markets due to advantages in terms of economies of scale and scope leading to lower transaction costs and improved bargaining power (World Bank, 2007: 154; Reardon and Barrett, 2000; Sexton and Iskow, 1988; Staatz, 1987; Granovetter, 1985).

---

[45] Known milk borne infectious diseases include typhoid fever, scarlet fever, septic sore throat diphtheria, tuberculosis, and brucellosis, among others. Unknown diseases can result from bacteria mutation or cross-contamination. (O'Connor, 1995).

[46] This information was obtained from: http://www.who.int/medacentre/factsheets/fs237/en/

[47] www.eufic.org

[48] www.fda.gov

Due to high demand pressure, and the need of smallholder farmers to maximise short-term profit to survive, cooperatives in developing countries may tend to develop structures and services to boost production and productivity, with little or no regard for quality. As quality specifications from the public sector and from supply chains become more stringent, the competitiveness of cooperatives in the market is increasingly challenged (World Bank, 2007: 156). Since agricultural cooperatives in developing countries are typically village-level organisations, owned and controlled by smallholders with limited managerial skills, they may not be able to face such a challenge (World Bank, 2007: 156).

The scope of this study is to measure the quality performance of Ethiopian dairy cooperatives and identify some incentives to improve it, given farm production and productivity and market specifications. Ethiopia is particularly suitable for this analysis being a country with high potential for dairy production, where both agricultural cooperatives and dairy value chains are growing rapidly (see Chapter Two and Four), and where quality performance of dairy cooperatives appears to be poor and largely overlooked (see Chapter Five). The following sections of this study define the analytical framework, describe the sample and data available, present an empirical model to interpret the data, discuss econometric results, and finally present conclusions and implications.

6.2 Analytical framework

Perishable commodities are usually subject to considerable quality variation and quality performance may easily become suboptimal due to both internal and external factors. Quality variability depends on the set of incentives faced by producers (Weaver and Kim, 2001). Blackman (2001) states that larger quality variability corresponds to high transaction costs and lack of trust between producers and buyers. The latter effect is mainly due to asymmetric information underlying variability in production management. Berti *et al.* (1998) take production capacity and economies of scale as major indicators for quality variation, whereas Teratanavat *et al.* (2005) relate quality performance to firm size and fixed investments. Quality performance can also be specified in terms of the techno-managerial performance (i.e. Luning *et al.*, 2006). In this approach, both the technical characteristics (production and processing) and the behavioural characteristics (information and exchange) are taken into account.

Production technology and management, have direct impact on observable and imminent dairy quality characteristics. Although a fair amount of knowledge is available regarding the technological options for improving milk quality (see Balasini, 2000; Taneja, 1999), it is far less understood which organisational form could be effective for providing farmers with the necessary incentives to enhance milk quality. In this chapter, we focus attention on the role of cooperative arrangements for quality management (a) as a possible cost-reducing device to enhance producers' compliance with market regulations and related grades and standards (Reardon and Barrett, 2000) and (b) as a strategy to control free-riding amongst producers and reinforce mutual trust and coordinated behaviour (Granovetter, 1985).

The first role of collective action is to provide farmers with an advantage in *economies of scale* and bargaining power. The second role of cooperation refers to aspects of improved control and coordination that should results in *economies of scope*. We focus attention on the latter aspect of (horizontal) coordination that reflects vertical supply chain interactions (e.g. price regimes, contracts and grades). Therefore we specify milk quality in cooperatives on the basis of mutual respect and inter-agency trust, usually enforced through asset-specific investment and quality control tools.

The analytical framework derived from these considerations explicitly focuses on cooperative organisation as a key agency in the promotion and compliance with milk quality standards. For practical purposes, reduced bacterial contamination and improved fat and protein content are considered as objective variables. These can be influenced by (a) collective organisation, expressed by the structure, grades and standards adopted by cooperatives, given (b) market-specific characteristics related to geographical location, and (c) farm-household characteristics, defined by age and gender of the farmer, herd size and animal phenotype. This approach draws from the framework elaborated by Saenz-Segura (2006) and Zuniga-Arias (2007) who express production quality on the basis of farm-household characteristics, institutional arrangements and price and non-price specifications from the market.

6.3 Data

The data-set available to this study provides a unique combination of biological, socio-economic and managerial information about dairy cooperatives in Ethiopia. The sample includes 10 primary dairy cooperatives selected from a list provided by the Ethiopian Federal Cooperative Commision (FCC).[49] Table 6.1 and Figure 6.1 show the spatial distribution and characteristics of the sample sites.

The 10 cooperatives were selected from the highland regions where milk production has more potential for development, as compared to the lowland regions (Ahmed *et al.*, 2003). The highlands occupies two-thirds of the country's territory, and are characterised by vast plateaus ranging from 1400 up to more than 3000 meters above sea level. The typical topography of the Ethiopian Highlands provides a suitable microclimate for the introduction of high-yielding dairy cows. Here, biophysical attributes, like the availability of vast grazing areas, mild slopes and fertile soil, adequate rainfalls patterns (1000-1900 mm/year) and temperature (0-30 °C) offer a relatively disease-free environment with high potential for animal feeding.

The sample is structured to represent the dairy cooperative network of the Ethiopian Highlands, as observed in 2005. The sample is diverse covering four regions and six zones.[50] The 10

---

[49] A primary dairy cooperative is defined as an association of farmers owning and controlling one centre for milk collection and inputs distribution. While secondary cooperatives are defined as unions of two or more primary cooperatives. FCC is the governmental organisation in charge of cooperative governace.

[50] Zones are below regional level and above municipality level.

Cooperation for competition

Table 6.1. Sample characteristics, Ethiopian Highlands, 2005.

Site (distance and direction from Addis Ababa)	Region & zone	Altitude (masl)[1]	N. of primary cooperatives	N. of farmers per cooperative	Farmers interviewed & milk samples taken
Dessie (400 km North-East)	Amhara South Wollo	2,700	1	45	15
Debre Zeit (50 km South)	Oromo South Shewa	1,600	4	33-62	10-20
Dejen (200 km North-West)	Amhara East Gojam	2,400	1	88	30
Selale (40 km North)	Oromo North Shewa	2,500	1	72	25
Selale (80 km North)	Oromo North Shewa	2,500	1	102	35
Asella (200 km South East)	Oromo Arsi	2,500	1	58	20
Harar (500 East)	Oromo-Somali Harerge	1,400	1	43	15

[1] Metres above sea level.

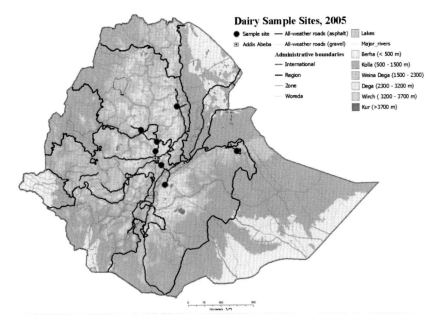

Figure 6.1. Sample sites, Ethiopian Highlands, 2005.
Source: IFPRI-ESSP.

cooperatives were selected in respect of the number and size of primary dairy cooperatives per region and zone (as reported by the FCC).

The number of farmers interviewed per cooperative was set in proportion to the overall number of members in each cooperative. Information was collected through direct interviews with cooperative managers (chairmen and members of the executive boards), as well as with member-farmers. Information from managers and farmers was collected using structured questionnaires. Farmers were interviewed at the milk collection centres, by picking every other farmer arrived to deliver its milk supply to the cooperative. Two small samples of milk were also collected from each interviewee.

All milk samples were delivered to the laboratory within 12 hours from collection, in order to analyse their microbiological and chemical attributes. Milk sampling followed the standard procedures applied by the International Livestock Research Institute (ILRI, CGIAR; defined by O'Connor, 1995), and the preservation of milk samples between collection and analysis was ensured using ice boxes. All milk samples were gathered and analysed within a one-month period, so as to reduce the influence of seasonal factors on milk attributes. Each milk sample was analysed for fat content (using the Gerber method), protein content (Protein Formaldehyde Titration method), and bacterial contamination (Plate Agar method), following standard practices applied by ILRI and specified by O'Connor (1995).

6.4 Ethiopian dairy cooperatives

The market of Addis Ababa is by far the largest, wealthiest and more diverse of Ethiopia. Six of the dairy cooperatives sampled, those located within 100 km from the capital (Table 6.1), sell directly to processing firms in Addis Ababa, while the other four sell in local markets. The average cooperative in our sample is 13 years old (established in 1993), and counts 59 member-farmers located within 10 km from the cooperative's milk collection centre, which is usually within or at the edge of urban areas. Milk supplies of member-farmers are collected on a daily basis, bulked, cooled in a few cases (only the cooperatives supplying Addis Ababa have cooling tanks), then processed into fermented butter, cottage cheese, sour milk or yoghurt, and finally marketed. Currently none of the cooperatives have machineries for milk sanitation (pasteurisation, UHT, filtration, etc.).

Before milk is collected, cooperative technicians screen each and every milk supply received on the basis of their sensorial perception, and/or by using simple field tests (71 percent of the cooperatives use the alcohol test, and 83 percent use the specific gravity test). [51] These techniques allow to evaluate milk quality on the spot, as good or bad, as opposed to more lenghty and careful evaluations based on a continuous scale of grades. Milk supplies that do not comply with the quality standards imposed by sensorial, alcohol and gravity tests are rejected, even if the rejection rate appears to be negligible. In most cases rejections contributes to raise internal arguments or even conflicts about the precision of quality control techniques and the integrity of the technicians.

The simple techniques for quality control adopted by dairy cooperatives reflects a national shortage of public and private laboratories and certification authorities, necessary to conduct a precise and transparent quality control of milk trade. In particular, the Quality Standard Authority, a major governmental body in charge of food quality control, has infrastructures and human resources that are far below the necessary capacity to enforce milk quality regulations in the national milk market. Private, third party certification authorities are completely absent.

Since milk quality control in cooperatives does not provide a range of quality grades, the price received by cooperative members cannot be set proportional to milk quality. It follows that milk price fluctuations in Ethiopia are mainly associated with variations in production, productivity and demand, rather than quality. On average the price received by cooperative members is estimated at 1.8 birr per liter of milk (approximately 0.16 Euro; Table 6.2), which would theoretically allow Ethiopian cooperative farmers to compete in most international markets (see Staal, 1995).[52]

---

[51] *The alcohol test* is a low cost technique for the instantaneous evaluation of milk colloidal suspension, overall hygiene and freshness (O'Connor, 1995). *The specific gravity test* is a low cost technique for the instantaneous evaluation of milk density, given milk temperature. This test is particularly useful to detect undeclared water addition and cream removal (O'Connor, 1995).

[52] Milk export from Ethiopia is negligible (Ahmed *et al.,* 2003).

Table 6.2. Major characteristics of dairy cooperatives, Ethiopian Highlands, 2005.

Variables (189 obs.)	Mean	Std. Dev.	Min.	Max.
Herd size (n. of milking cows)	2.1	1.2	1	7
Dummy for crossbred herd	0.78	0.41	0	1
Production (lt/farm/day)	11.6	11.5	1	73
Productivity (lt/day/cow)	6.1	4.2	0.25	23
Price received by farmers (Birr/lt)	1.8	0.7	1.13	3.75
Farmer age	46	12.2	19	80
Dummy for female farmer	0.33	0.47	0	1
Farm-coop distance (km)	2.6	2.8	0.001	10
Total bacterial count of milk (cfu/ml)[1]	608 million	2.34e+09	200	1.0e+10
Fat content in milk (%)	4.0	0.9	2	9
Protein content in milk (%)	3.0	0.3	2	4.18

[1] Colony forming units.

On the other hand, quality and safety attributes of milk from Ethiopian cooperative farmers could theoretically impede milk export from Ethiopia. The milk quality analyses conducted during the survey period reveal that the milk produced by Ethiopian cooperative farmers has higher bacterial contamination and lower protein and fat content, compared to all standards and secondary data available from both developing and developed countries (Table 6.2; Figures 6.2 and 6.3).

The majority of cooperative members are males (67 percent) with an average age of 46 (Table 6.2). Cooperative farmers have preferential access to artificial insemination services, which are fully subsidised and managed by the state. As a result, 78 percent of cooperatives' cows are crossbred (presenting mixed Zebu and Frisian phenotypes), while the rest are pure indigenous Zebus (Table 6.2).[53] Cross-breeding activities through artificial insemination have the objective to expand herds and yields, which however appear still rather small, given an average of two cows and 12 liters of milk per day per farm. Potential explanations to poor productive performance include limited coverage and quality of artificial insemination services, and limited availability of training and information on husbandry and feeding practices (Ahmed *et al.*, 2003).[54]

---

[53] Ethiopian cooperative farmers have heterogeneous herds, composed by zebu cows and/or crossbred cows. The latter are hybrid genotypes, usually characterised by different proportions of Frisian and Zebu genes, and by higher milk productivity compared to pure indigenous zebu cattle.

[54] Crossbred herds are typically kept inside the barn in Ethiopia. Grazing would mean to expose valuable and less resistant crossbred cows to an environment to which they are not naturally adapted. In the barn crossbred cows are commonly fed with hay and crop residues, seldom concentrated feed.

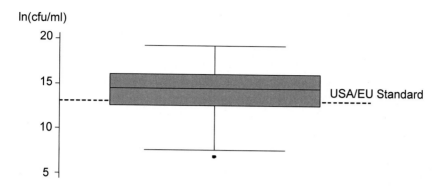

Figure 6.2. *Total bacterial count in milk collected from cooperative farms, Ethiopian Highlands, 2005.*

Source: *Household data collected by the author. Total Bacterial Count (TBC) is measured counting the number of bacterial colonies in one ml of milk (cfu/ml). The measurements units in the graph are expressed in logarithms in order to reduce the large variability observed in the sample. The dotted line included in the graph corresponds to the public standards adopted by US and EU (see EUFIC (European Food Information Council): www.eufic.org and FDA (Food and Drug Administration): www. fda.gov) for raw milk (2 million cfu/ml or 14.5 ln (cfu/ml)).*

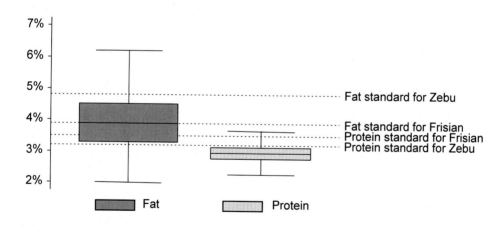

Figure 6.3. *Fat and protein content in milk collected from cooperative farms, Ethiopian Highlands, 2005.*

Source: *Household data collected by the author. Fat and protein standards included in the graph are an average of secondary data from both developing and developed countries world-wide (Taneja and Aiumlamai, 1999; Walstra et al., 2006). Fat standards are set at 4.8 percent for zebu cattle and at 3.9 percent for Frisian cows. Protein standards are set at 3.2 percent for zebu cattle and 3.5 percent for Frisian cows.*

6.5 Empirical model

In the proposed model for the case of Ethiopian dairy cooperatives, milk quality standards, are given by two dummies indicating alcohol, *al*, and specific gravity tests, *sg*. When the equipment needed for these tests is not available cooperative technicians rely on their sensorial perception to screen the milk supplies delivered by member-farmers. Quality standards are crucial to minimise subjectivity in quality control. However, Weaver and Kim (2001) argue that tests like those adopted by Ethiopian cooperatives can turn into disincentives for quality whenever the standards imposed by the tests lie below the actual quality of production. Price premiums for milk quality are not included in the empirical model since the tests available to Ethiopian cooperatives for quality control do not allow for milk quality grading on the basis of a continuous scale of values, and therefore milk quality cannot be priced.

As suggested by Tegegne (2003), the organisation of Ethiopian dairy cooperatives depends also on membership structure, and in particular on the number of members and the distance between them and the cooperative centre used for milk collection, and inputs and information distribution. In particular, Tegegne (2003) suggests that access to input and output markets improves with bargaining power, which is greater for large cooperatives, as well as with the efficiency of the distribution systems, which is expected to be greater in small cooperatives. For these reasons the relationship between cooperative size and performance is expected to be exponential, describing a concave curve, which can be captured introducing cooperative size, *cs*, and its squared term in the empirical model.

Besides, as cooperative centres for milk collection and input distribution are typically found within urban areas, farm households that are farther located from these centres may have better access to pastures, favouring livestock free range and feed diversification. Farms located in or in proximity of cooperative centres and urban areas may have space constraints and limited access to rural pastures, favouring permanent in-barn (indoor) husbandry. Since barns in Ethiopia are typically small and poorly constructed they affect animal welfare. Moreover, since animal feed for in-barn feeding is extremely scarce and of poor nutritional value in Ethiopia (Ayatunde *et al.*, 2005), urban farmers are expected to produce milk of lower quality. For these reasons the empirical model includes a variable measuring the distance (in km) between farms and the collection centre, *d*. Overall, collective services might depend also on the experience accumulated by a cooperative over time, *ca*. Learning by doing is particularly important in cooperatives in developing countries, due to poor managerial knowledge among member-farmers and shortage of capital to hire professional managers (World Bank, 2007: 156).

The impact of collective organisation on milk quality is conditional on market incentives for production. Weaver and Kim (2001), suggest that different markets provide different incentives for milk quality at the farm gate. In the context of Ethiopia, dairy cooperatives that have direct access to the market of Addis Ababa, by far the largest and most developed of the country, are therefore expected to face different demands and preferences compared to cooperatives that sell in and procure from other national markets. Consequently, in this study we capture variations in market-specific preferences with a dummy indicating cooperatives with direct access to the market of Addis Ababa, m.

Milk quality at the farm gate depends also on farm-specific production technology. Age, fa, and gender (dummy for female), g, of member-farmers can capture some of the differences in farming ability and motivations, while the size of members' herds (measured in number of milking cows), h, and a dummy indicating herds dominated by high yielding crossbred cows (rather than indigenous zebus), y, can capture farm-specific asset. In particular, we expect that larger herds can benefits from economies of scale in procuring feed, and therefore produce milk of better quality. However, larger herds can also have a negative effect on milk quality. For instance, in farms where land is scarce and barns are small the expansion of herds could lead to a reduction in animal welfare. Finally herds dominated by crossbred cows are expected to produce more milk characterised by lower nutrient density, compared to herds dominated by indigenous zebu cattle (as extensively documented in dairy science, see Balasini, 2000; and Taneja, 1999).

The resulting empirical model (Equations 6.1, 6.2, 6.3), explains total bacterial count, tbc, fat and protein content, fat and $prot$, in the milk produced by a cooperative farmer i:

$$tbc_i = \beta_0 + \beta_1 al_i + \beta_2 sg_i + \beta_3 ca_i + \beta_4 h_i + \beta_5 y_i + \beta_6 m_i + \beta_7 d_i + \beta_8 cs_i + \beta_9 cs_i^2 + \beta_{10} fa_i + \beta_{11} g_i + e_i \quad (6.1)$$

$$fat_i = \beta_0 + \beta_1 al_i + \beta_2 sg_i + \beta_3 ca_i + \beta_4 h_i + \beta_5 y_i + \beta_6 m_i + \beta_7 d_i + \beta_8 cs_i + \beta_9 cs_i^2 + \beta_{10} fa_i + \beta_{11} g_i + e_i \quad (6.2)$$

$$prot_i = \beta_0 + \beta_1 al_i + \beta_2 sg_i + \beta_3 ca_i + \beta_4 h_i + \beta_5 y_i + \beta_6 m_i + \beta_7 d_i + \beta_8 cs_i + \beta_9 cs_i^2 + \beta_{10} fa_i + \beta_{11} g_i + e_i \quad (6.3)$$

The model could suffer from econometric problems that need to be addressed before discussing the results. First of all, since fat and protein content are simultaneously determined within the mammary cell, and statistically correlated (55 percent correlation), disturbances to fat and protein synthesis may coincide, justifying an estimation based on Seemingly Unrelated Regression (SURE).

Second, with econometric models that are built on data collected at one point in time, as in this case, it is difficult to ascertain that right hand side variables cause variations of the dependent variable, rather than the other way around (endogeneity). However, milk quality grades, as measured in this study, are not observable in reality, due to the lack of milk grading techniques at the farm gate. Consequently, it is unlikely that milk quality grades influence the choice of cooperative conduct and structure, given the explanatory variables specified in the model. Additional estimation problems are related to potential biases due to omitted variables, heteroskedasticity or not normally distributed residuals. Bias due to omitted variables was tested and excluded using the Ramsey regression specification-error test for omitted variables (default in STATA). Skewness and kurtosis tests indicate that residuals are normally distributed at five percent significance level, while heteroskedasticity was controlled by estimating robust standard errors (not in the SURE model), and applying logarithmic functional forms.

6.6 Results

Empirical findings (Table 6.3) suggest that an additional year of cooperative experience, contributes to reduce bacterial contamination (by 372 cfu/ml). Milk hygiene improves with experience and with increasing exposition to cooperative services, which often include advice and in some cases also training on milking and milk handling practices. Milk bacterial contamination decreases (by 23 cfu/ml) also when cooperatives have direct access to the market of Addis Ababa.

On the other hand, direct access to the largest and most developed market of Ethiopia reduces milk fat content by 1.1 percent. These findings suggest that the market of Addis Ababa provides better incentives to improve milk hygiene, but disincentives to improve fat content compared to other markets in the Ethiopian Highlands. A 1.1 percent reduction in fat content is observed also in herds dominated by cows with crossbred phenotype. As expected, crossbreeding activities in Ethiopia favour productivity to the detriment of milk nutrient density.

Table 6.3. Determinants of milk quality at the farm gate (OLS), Ethiopian Highlands, 2005.

Explanatory variables	Dependent variables		
	(ln) Fat content	(ln) Protein content	(ln) Total bacterial count
Market characteristics			
Access to Addis Ababa market (dummy)	-0.09 (0.05)*	0.03 (0.02)	-3.14 (0.59)**
Cooperative characteristics			
Specific gravity test (dummy)	-0.98 (0.34)**	-0.42 (0.15)**	1.07 (3.56)
Alcohol test (dummy)	-0.16 (0.11)	-0.07 (0.05)	-0.88 (1.40)
(ln) Cooperative age	-0.10 (0.23)	0.05 (0.11)	-5.92 (2.57)**
(ln) Coop-size	7.18 (2.18)**	2.11 (1.00)**	18.4 (25.4)
(ln) Coop size square	-0.88 (0.27)**	-0.25 (0.12)**	-2.35 (2.11)
(ln) Farm-coop distance (Km)	0.04 (0.03)	0.03 (0.01)**	0.64 (0.41)
Farm household characteristics			
(ln) Farmer age	0.04 (0.06)	0.01 (0.03)	1.14 (0.64)
Female farmer (dummy)	-0.02 (0.03)	0.01 (0.02)	-0.06 (0.38)
(ln) Herd size	- 0.01 (0.03)	- 0.00 (0.01)	- 0.45 (0.34)
Crossbred herds (dummy)	- 0.08 (0.05)*	0.01 (0.02)	0.34 (0.55)
R-squared	0.3288	0.2402	0.2563
N. of observations	185		

Standard error in parentheses (), *denotes significance at 10% level, **denotes significance at 5% level.

Both milk fat and protein content increases with cooperative size and decreases with its squared term. The optimal cooperative size for milk quality is estimated at 60 members.[55] Smaller cooperatives are likely to have less bargaining power vis-à-vis governmental and non- governmental organisations (especially for procuring training on animal husbandry and feeding), and vis-à-vis the market (especially to procure concentrated feed). Larger cooperatives (with more than 60 members) may instead put an already limited managerial capacity under further stress.

Results suggest also that when the distance between cooperative farms and the cooperative centre (for milk collection, inputs and information distribution) increases by one km, milk protein content increases by one percent. The large majority of collection and distribution centres of cooperatives are located in urban areas, where land is scarce by definition. Hence, farms located close to the cooperative centre are expected to have limited space for the herd, and difficult access to pastures, reducing the welfare of the cows, and the bio-diversity of their feed intake.

Finally, the techniques adopted by cooperatives to screen members' milk (alcohol and specific gravity test) appear to set quality standards that are either less effective or not significantly different compared to traditional techniques based on sensorial perceptions. While the alcohol test has no significant impact on milk quality at the farm gate, the use of specific gravity test implies a drastic reduction (by 2.7 percent) in milk fat content, as well as in protein content (by 1.5 percent). These results are particularly striking. However, cooperative managers' report that the incidence of milk supplies that do not comply with alcohol and gravity tests is negligible, suggesting that the standards associated with these tests lie well below actual quality of milk supplies. Moreover, direct observation of quality control practices reveal that the specific gravity test is commonly misused.[56]

6.7 Conclusions and implications

This study provides some recommendations on how to build and manage dairy cooperatives in Ethiopia to improve the quality and safety of the milk produced by smallholders, so that the latter can better compete in the marketplace. First of all this study shows that Ethiopian dairy cooperative produce milk characterised by lower fat and protein content, and higher bacterial contamination, compared to European and US standards (see EUFIC (European Food Information Council): www.eufic.org and FDA (Food and Drug Administration): www.fda.gov and Walstra *et al.*, 2006), and evidence available from the tropics (Taneja, 1999). Econometric results suggest that given market-specific characteristics, and farm-specific

---

[55] The optimal cooperative size is computed dividing the coefficient for the number of members, β_8 by the coefficient for the squared number of members (in absolute value), β_9. The resulting value is divided by two and then applied as exponent to the base e: $e^{[(\beta8/\beta9)/2]}$.

[56] The specific gravity test is usually applied without measuring milk temperature. Since milk density varies according to milk temperature, the results of the specific gravity are unreliable.

production asset, milk quality and safety could be still improved rearranging structure, services, and grades and standards adopted by cooperatives.

Farmers' access to concentrated feed, training and advice on quality management depends on bargaining power, which is greater for large cooperatives, but it also depends on procurement and distribution systems, which are more efficient in small cooperatives. Empirical findings suggest that milk quality could be improved with the optimisation of cooperative size to approximately 60 members per milk collection centre. As observed in the cooperatives of *Selale* and Debre Zeit, and suggested by Tegegne (2003), the optimisation of cooperatives' size, could involve joint-ventures between small cooperatives (union of two or more cooperatives), or the partitioning of big cooperatives through the establishment of additional centres to collect milk, and exchange information and inputs. These centres should be gradually moved out of urban areas, where they are commonly found, and placed closer to grazing areas whilst ensuring good connection with major urban markets. In this way dairy cooperatives could promote mixed feeding systems, based on both grazing and in barn feeding, and discourage unhealthy urban farming in which cows are constantly kept inside barns, in close contact with human settlements.

Moreover, results suggest that milk hygiene improves as cooperatives get older and increase the exposition of members to information and training to improve the barn hygiene, milking and milk handling practices. Finally, Ethiopian dairy cooperatives need to improve the quality control techniques adopted to screen and regulate milk flow from farmers to the market. As suggested by Weaver and Kim (2001), tests measuring milk quality and safety as good or bad, need to be replaced with techniques that allow quality evaluations based on a continuous scale of grades. Binary tests measuring milk as good or bad can introduce quality standards that lie below the actual quality of milk supplies creating disincentives to milk quality upgrading (as in the case of Ethiopian Dairy Cooperatives). On the contrary, tests based on continuous grading scales would avoid the latter problem, and allow to set progressive standards and price premiums for continuous quality upgrading.

Due to the fragile willigness of Ethiopian consumers to pay for milk quality, and the shortage of capital and managerial expertise among smallholder farmers, dairy cooperatives need external support to reorganise. Given the importance of milk quality and safety for public health, processing and retail efficiency, external incentives for quality should come especially from both public institutions and private corporations (i.e. industries and supermarkets). Further research and roundtables are needed to define public, industrial, retail and cooperatives' responsibility.

'Co-operatives are an instrument for the non-conflicting incorporation of the proletariat into economic development.'
Luigi Luzzatti, Pollicy Maker for the Liberal Government of Giovanni Giolitti, 1908, Italy.

Chapter 7. Conclusions and implications

7.1 Key debate

'Global competition scares smallholders away from the market, public support is shrinking or inefficiently governed, economists fail to provide incentives for small and micro business development, consequently smallholders rediscover the importance of the community or collectivity' (Di Vico, 2008). Collective action is a potential instrument for smallholders to cope with the challenges posed by global markets. Under communist regimes (including the Derg in Ethiopia), collective action has often been turned into an instrument for political elites to patronise the aspirations of communities in order to perpetuate mainstream ideologies. Today, 'communalisation' or 'collectivisation' has returned to offer the services that the state is abandoning (Bauman and Tester, 2001). In particular, many donors and governments consider agricultural cooperatives a fundamental pillar of their rural development policy, as well as a core institution in the process of governance decentralisation and agri-business development (World Bank, 2007: 155). As demonstrated by the growing number of rural cooperatives in the developing world, community empowerment and collective entrepreneurship appear to be particularly viable solutions, especially where infrastructures and markets are poorly developed.

As Rorty (1991) put it, *'globalisation is producing a world economy...that will soon be owned by a cosmopolitan upper class which has no more sense of community with any workers anywhere than the great American capitalists of the year 1900 had with the immigrants who manned their enterprise'*. Community empowerment and collective action are frequently advocated in response to these concerns.

Certainly agricultural cooperatives are advocated by the Ethiopian government (FDRE, 1994, 1998, 2002, 2005), which set the ambitious target to promote the formation of at least one cooperative per *kebele* to 70 percent of the national *kebeles* by 2010 (FCC, 2006).[57] As a result the share of *kebeles* with cooperatives went up from 10 percent in 1991 to nearly 35 percent in 2005 (Bernard *et al.*, 2008). In particular, in 2005, nine percent of the rural households (comprising approximately 6 million individuals) were engaged in some form of cooperatives in Ethiopia. Since 85 percent of the Ethiopian population lives in rural areas under subsistence or semi-subsistence regimes (Alemu *et al.*, 2006; CSA, 2000), agricultural

---

[57] In Ethiopia the *kebele* is the smallest administrative units, below the municipality-district level.

cooperatives are seen as key institutions to link national smallholder farmers to emerging supply chains and commodity exchange networks. In an era of growing socio-economic disparity, horizontal organisational forms, based on pooling and bonding mechanisms, are expected to facilitate vertical integration between smallholders and the market. In particular, agricultural cooperatives are envisaged as favourable organisational forms to improve bargaining power and reduce the transaction costs of smallholders (Helmberger and Hoos, 1995; Nourse, 1945), in markets where power is retained by retail/industrial oligopolies/oligopsonies and information is highly asymmetric across buyer and seller (Staatz, 1983; Sexton, 1986, 1988).

However, substantial part of the literature suggests that agricultural cooperatives are not always successful business organisations (see Damiani, 2000; Neven *et al.*, 2005; Chirwa *et al.*, 2005; Sharma and Gulati, 2003; Uphoff, 1993; Attwood and Baviskar, 1987; Tendler, 1983). Agribusiness literature emphasises the complexities added when multiple individuals, rather than a single investor, engage in commercial activities (Cook and Chambers, 2007; Cook, 1995; Putterman and DiGiorgio, 1985; Vitaliano, 1983; Fama, 1980; Jensen and Meckling, 1976; Olson, 1965). In developing countries, cooperatives are typically village-level, community-based, organisations that face considerable difficulties in combining social equity purposes with commercial activities (World Bank, 2007: 155). Furthermore, as cooperatives have been largely government controlled and staffed, they are often considered as an extended arm of the public sector, rather than institutions or firms owned by the farmers. This form of cooperatives has been rarely successful. Political patronage and interference has too often resulted in poor business performance, corruption and conflicts, which have contributed to discredit the movement. Based on similar arguments, Bernard *et al.* (2008) suggest that the formation of cooperatives provides no clear advantages for the commercialisation of grains in Ethiopia.

In the previous chapters we presented a detailed analysis of the cooperative movement in rural Ethiopia. This final chapter summarises the main findings of this study in response to the research questions presented in Chapter One. Consequently, it elaborates further on the theoretical contribution of this book to the current debate and literature on food markets and rural cooperation. Finally, it elaborates on the implications for policy and for further research discussed in the previous chapters.

7.2 Main findings

The analysis of the cooperative movement in rural Ethiopia starts from the analysis of external conditions dictated by changes in the agri-food market:

a. What are the trends and challenges in the Ethiopian agri-food markets?

This study identifies three major trends in the Ethiopian agri-food market, in line with the scenario described in most developing countries (World Bank, 2007: 118-120) (1) increasing market concentration into supermarket retail formats (see Chapter Three), (2) increasing market integration into value adding chains for perishable agri-food supplies (see Chapter Three), and (3) increasing market organisation into competitive and coordinated exchange networks for storable agri-food commodities (see Chapter Two). These trends pose significant challenges to Ethiopian smallholders. Major challenges arise due to market power concentration into retail/industrial oligopolies and oligopsonies, and the limitations of public infrastructures (roads, telecommunication, electricity, water supply, etc.). Market power concentration by supermarkets and processing industries results in more hygienic products with longer shelf life, and more hygienic and convenient (time-saving) outlets. However, it also translates into reduced outlets' accessibility, since supermarkets tend to concentrate in wealthy urban neighbourhoods; the retail concentration process is associated with an increase in the distance between market outlets and poor households. Due to oligopoly power, the preferences of consumers, and especially of poor consumers, are more likely to be ignored in the process to design outlets and products. Due to oligopsony power, price, quality and safety specifications tend to become more stringent favouring capital intensive producers over smallholder farmers. While retail/industrial oligopolies and oligopsonies threaten to exclude smallholder consumers and farmers from emerging value chains, asymmetric information across buyers and sellers and high transportation costs (related to poorly developed infrastructures) tend to keep smallholders far away from agri-commodity exchange networks. Information and delivery problems are of particular importance in Ethiopia, where 85 percent of the population lives in depressed rural areas, mainly producing and consuming staple cereals, such as wheat, maize, teff, or traditional commodities such as coffee and oil seeds.

b. What is the impact of collective action for the competitiveness of Ethiopian rural smallholders?

This study suggests that the impact of collective action on the competitiveness of Ethiopian smallholder farmers is a controversial issue. First of all, Ethiopian agricultural cooperatives tend to exclude the 'poorest of the poor' (i.e. rural households with little or no land; see Chapter Two). Second, the average agricultural cooperative facilitates farmers' access to subsidised inputs (mainly fertilizer, improved seeds and artificial insemination), but does not have a significant impact on promoting farm output commercialisation see (see Chapter Two). Collective action promotes agricultural commercialisation only when it creates access to alternative and more profitable market outlets (see Chapter Two). In Ethiopia, almost half of the existing cooperatives do not create this opportunity, so that commercialisation remains dependent on individual entrepreneurship and resources (i.e. commercialisation outside the cooperative). These cooperatives overshadow the impact of the other half of the Ethiopian cooperatives that instead contribute to promote farm output commercialisation on the basis of collective marketing mechanisms (i.e. commercialisation through cooperatives). Third, when collective action embraces collective marketing, farmers tend to intensify production (volumes and productivity) to the detriment of output quality and safety (see Chapter Four). Collective marketing activities provide clear incentives to intensify production. However, due to high demand pressure for cheap food and widespread poverty among producers, collective marketing tends to neglect the importance of output quality and safety in Ethiopia. Because of extremely poor quality and hygiene of production, Ethiopian cooperatives are highly vulnerable to competition and susceptible to bans from (inter)national markets.

c. How to promote the competitiveness of Ethiopian rural cooperatives?

Based on previous findings, this study identifies the need to expand collective marketing activities, and to improve production quality and safety, as top priorities in the process to promote the competitiveness of Ethiopian agri-cooperatives. In Ethiopia, collective marketing activities are more likely to occur in cooperatives established upon the voluntary initiative of farmers, under conducive market and governance conditions (see Chapter Six). Markets and governance vary across Ethiopia. Ethiopia is a federated Republic where regions have a semi-independent status. The northern regions of Amhara and Tigray have a longer history of trade, and are more advanced in terms of infrastructures and urbanisation compared to the rest of the country. Regional disparity in Ethiopia is the likely result of political clientelism throughout history. While Tigray is the homeland of the current ruling party, which has been in power since 1991, the Amharas were the ethnic group that dominated the country during the longstanding empire (1930-1974) of Haile Sellaise (himself an Amhara). For these reasons, market and governance conditions in the southern regions tend to be less conducive for collective marketing than in the northern regions.

Interventions by governmental and non-governmental agencies to promote the formation of agricultural cooperatives are often too invasive in Ethiopia, creating rural dependency rather than entrepreneurship. As observed in many other developing countries (World Bank, 2007: 156), cooperatives in Ethiopia appear to be too often used as instruments to implement policies designed without consulting them, in order to fulfil government's and donors' agenda. Top-down interventions tend to attract opportunistic and subsistence farmers, eager to extract subsidies rather than embark in marketing activities. On the contrary, cooperatives founded on the spontaneous initiative of farmers are more likely to develop sustainable commercial activities. As put by Cook and Chambers (2007), collective marketing faces cyclical challenges. The business cycle of cooperatives is characterised by an initial stage with high turnover, followed by a reduction in sales due to increasing competition. After establishment, Ethiopian cooperatives are found to enter a period of growing competitiveness, which reaches its peak after approximately eight years. Subsequently, collective competitiveness begins to diminish. This study shows that competitiveness fades remarkably faster in cooperatives established on external (top-down) initiatives. While these cooperatives tend to embark on collective marketing early on, their competitiveness fades rapidly over time. By contrast, cooperatives established on bottom-up initiatives develop collective marketing activities gradually over time, and although they face cyclical challenges in the marketplace, their competitiveness tends to diminish at a much slower pace.

To reduce their vulnerability to market competition, Ethiopian cooperatives need to improve the quality of their production. To upgrade production quality, minimizing drawbacks in terms of production quantity and productivity, Ethiopian cooperatives have two main options (see Chapter Five): (1) to improve quality control at the farm gate, and (2) to improve farmers' access to land and to information on quality management. Most Ethiopian cooperatives screen the quality of farmers' supplies on the basis of sensorial perception, and/or using field tests that measures quality as good or bad (rather than on a continuous scale of grades). Quality control techniques of the type adopted by Ethiopian cooperatives can be counterproductive. First of all, they do not provide precise and reliable information about product quality. Second, they do not allow setting progressive premium prices to ensure continuous quality upgrading. Consequently, they reduce heterogeneity in nutritional value, hygiene, and taste and level this quality attributes to a fixed standard. Since the share of farmers' supplies that are rejected is typically small in Ethiopian cooperatives, such a fixed standard is expected to lie below the actual quality of supplies, creating a disincentive for on-farm quality management. Third, quality control in Ethiopian cooperatives is typically conducted in the absence of independent arbitrage, i.e. in the absence of a third party (public or private institutions, or even anti-trust cooperative bodies) to certify the legitimacy of milk evaluation techniques. This shortfall allows for manipulation of buyers and consumers (meaning that supplies that do not comply with quality standards are nonetheless accepted and commercialised by the cooperative), or even of farmers (when good supplies are rejected), especially against those farmers that are disliked by the rest of the cooperative or simply by the technician in charge of quality control.

Increasingly stringent grades and standards need to be accompanied by improved access to land and knowledge about quality farming practices. Ethiopian cooperatives are typically organised around a centre, located in or in proximity of urban areas, where farmers deliver their output and procure information and inputs. These centres should be gradually moved out of cities and towns, whilst ensuring access to the market. By doing so Ethiopian cooperatives encourage the participation of farmers from remote rural areas, where land is cheaper and agricultural intensification can be achieved through better quality practices (e.g. by increasing soil fertility through pasture-crop rotations, or by adopting livestock free range instead of in barn husbandry). This study suggests that production quality and safety improves also through the optimisation of cooperative size. The capacity to procure information on quality farming practices at lower costs depends on bargaining power, which is greater for large cooperatives. However, the re-distribution of information to cooperative farmers is more efficient in small cooperatives. Empirical findings suggest that production quality is highest in primary cooperatives that count approximately 60 members.

7.3 Implications for research

In an increasingly globalised world, research on economic development of smallholders in Africa can no longer afford to limit itself to optimisation of rural livelihood support strategies and agricultural technology. In developed countries, improved competitiveness has long been one of the driving forces in agri-business research. This study contributes to current research on agribusiness for development by identifying possibilities for improving the competitiveness of smallholder farmers in Ethiopia. The resulting theoretical contribution of this study is depicted in Figure 7.1.

In Figure 7.1, farmer competitiveness is at its maximum when all agricultural output is sold in the market and when productivity and production quality are balanced, i.e. in correspondence of γ. By contrast, farmer competitiveness is at its minimum when agricultural output is not sold, but entirely consumed within the farm household, and when there is no balance in production between quality and quantity, i.e. in correspondence of α or β.

The trade-off between commercialisation and autarkic (or subsistence) behaviour is straightforward, in the sense that production strategies can respond to external demands from the market, or intra-household consumption needs. While commercialisation is associated with agricultural specialisation and a competitive advantage, autarky is typically associated with production diversification and missing market opportunities. Competitiveness depends

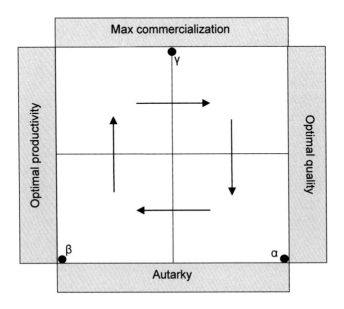

Figure 7.1. Patterns and determinants of farmer competitiveness in Ethiopia.

also on the trade-off between productivity and production quality. This trade-off reflects the 'intensification versus extensification' and/or 'indigenous versus exotic' dilemmas. For example, in Ethiopia (exotic) Frisian cows produce more milk than (indigenous) Zebu cattle, however, higher milk productivity is associated with lower nutrient density. Frisian cows are also more susceptible to diseases (especially mastitis) compared to indigenous cattle which are better adapted to the Ethiopian agro-ecology. Similarly, cows kept constantly inside barns under intensive feeding regimes (based on concentrated feed) produce more milk, but are also more susceptible to common diseases like mastitis, as well as brand new ones like the 'mad cow disease' (i.e. the Bovine Spongiform Encephalopathy, BSE). Extensive farming practices, based on free range livestock husbandry are far less productive, but also perceived as 'more natural' and healthy.

Overall, it is perceived that the intensification of livestock production systems lead to the worsening of animal welfare, which can in turn affect the quality of milk, meat, or eggs. To a large extent these dilemmas apply also to crop production. The use of chemicals is particularly important to grow crop varieties that are less adapted to the environment (i.e. exotic). The intensive use of inorganic fertilizer and pesticides leads to higher yields, but also higher health risk for consumers (and also for producers). Indeed, GMOs could offer a solution to minimise the productivity-quality trade-off, however scientific evidence appears to be still incomplete in this case, and the rise of ethical, social and environmental concerns, more than traditional quality concerns, seem to produce alternative trade-offs.

In Figure 7.1, the typical smallholder farmer in Ethiopia is found in the bottom-right quadrant, in proximity to α. This farmer produces mainly for subsistence using a traditional technology which results in little output with a relatively high-quality. This study shows that collective action enables Ethiopian farmers to move horizontally to the bottom-left quadrant, towards β, improving productivity to the detriment of quality, but maintaining a predominantly autarkic behaviour. When collective action involves collective marketing, farmers are able to move up, into the left-top quadrant, which includes farmers characterised by higher productivity and predominantly commercial behaviour. The findings of this study reveal the steps that marketing groups or cooperatives should follow for their members to get closer to point γ and remain there as long as the natural business cycle permits.

Not detracting from the important contribution that this study makes to current research on agri-business in developing countries, there are a number of issues that still remain to be addressed. In terms of follow-up, although this study has shown how the cooperative life-cycle can influence the competitiveness of farmers in rural Ethiopia, it does not provide a clear typology of the circumstances under which farmers operate at each stage of the cycle. Furthermore, a number of interesting directions can be suggested here to broaden the scope of the current study. First, marketing cooperatives are not the only form of farmer organisation or rural institution that can enhance smallholder competitiveness in supply chains and commodity exchange networks. Cooperatives generally emerge under conducive governance to address power concentration and asymmetric information in the market. Under different governance and market circumstances, alternative organisational forms of rural business can emerge. Vertical integration based on contract farming or off-farm employment could provide an alternative or complementary strategy to improve farmers' competitiveness. Second, to further support smallholder competitiveness, the role of institutions that can complement cooperatives, such as mechanisms to secure property rights, credit and saving institutions, weather-indexed insurance and institutional innovation for input markets, can and should be simultaneously explored. Finally, cooperatives are a very heterogeneous form of organisation differing both across time and space. Results of this study hold for agricultural cooperatives in the Ethiopian Highlands, research in different settings would contribute to a more general specification of the role of cooperatives in agribusiness.

7.4 Implications for policy

Ethiopia, which remains one of the poorest countries in the world, is undergoing a market transition with particular implications for agriculture, which forms the main source of subsistence for the large majority of the population. Ethiopia's favourable agro-climatic conditions mean that it has large untapped potential for agricultural growth and poverty alleviation. This study analyses cooperative organisations and the role they can play in stimulating farmer competitiveness vis-à-vis supply chains and commodity exchange networks. Ethiopia is currently witnessing a return to rural collective action. This can be interpreted as a community-level response to the extraterritorial challenges of a global market. Productivity and quality are becoming more important for Ethiopian smallholders to compete in an increasingly competitive market. Farmers face similar challenges everywhere; however, their response is bounded by local, community-specific constraints. Such constraints often result in collective action.

Not all types of cooperatives in Ethiopia appear to be viable institutions to promote commercialisation. Only those that collect and market their members' output are actually found to help smallholders to participate in the market. This implies that if the aim is to improve rural competitiveness, policy makers should selectively support such collective marketing activities. This approach could countervail the increasing concentration of power in supply chains and facilitate smallholders' participation in increasingly organised commodity exchange networks. To promote marketing cooperatives, public support should not impede with farmers' initiatives to organise collectively, but formulate appropriate support instead in the form of managerial capacity building.

When marketing cooperatives arise, their members tend to improve productivity but neglect quality. This reflects the policy focus on productivity increase through the provision of subsidies. Additional interventions are needed to improve quality and safety of agricultural output. In particular, quality control in Ethiopian cooperatives should move from binary to more detailed evaluations and from field tests to certified laboratory analyses.

The enforcement of better grades and standards should result from public-private partnerships where the role of the public sector is to provide arbitrage, and the role of the private sector (industries and supermarkets) is to provide incentives. Arbitrage requires a strong and capillary presence of the state in monitoring the quality in agricultural trade. Private incentives could come in the form of strategic alliances, or self-enforcing contracts between industries or supermarkets and cooperatives. Often such alliances do not arise due to power asymmetries in the market. Policy makers should encourage these alliances through facilitating the negotiation process and raising awareness of corporate social responsibility.

References

Ahmed, M.A.M., S. Ehui and A. Yemesrach, 2003. Dairy Development in Ethiopia. ILRI Working Paper no. 58, Addis Ababa, Ethiopia.

Alarcon, A., 2003. Ecuador Retail Food Sector Report 2003. USDA GAIN Report, December.

Alemu, D.E., E. Gabre-Madhin and S. Dejene, 2006. From Farmer to Market and Market to Farmer: Characterizing Smallholder Commercialization in Ethiopia. ESSP Working Paper. Washington, D.C.: International Food Policy Research Institute.

Alemu, D.E., and J. Pender, 2007. Determinants of Food Crop Commercialization in Ethiopia. International Food Policy Research Institute, Washington DC, Mimeo.

Attwood, D., and B. Baviskar, 1987. Why Do Some Co-operatives Work but not Others? A Comparative Analysis of Sugar Co-operatives in India. Economic and Political Weekly, 22 (26): A38-45.

Ayantunde, A.A., S. Fernandez-Rivera and G. McCrabb, 2005. Coping with Feed Scarcity in Smallholder Livestock Systems in Developing Countries. Wageningen University and Research Centre (The Netherlands), Animal Sciences Group, University of Reading (UK), Swiss Federal Institute of Technology (Switzerland), ILRI (Kenya).

Balasini, A., 2000. Zootecnica Speciale. Edited by Calderini Edagricole.

Balsevich, F., 2005. Essays on Producers' Participation, Access and Response to the Changing Nature of Dynamic Markets in Nicaragua and Costa Rica. Ph.D Dissertation, East Lansing, Michigan State University, USA.

Barham, B.L. and M. Childress, 1992. Membership Desertion as an Adjustment Process in Honduran Land Reform Cooperatives. Economic Development and Cultural Change, 40 (3): 578-614.

Bauman, Z. and K. Tester, 2001. Conversations with Zygmunt Bauman. Cambridge: Polity Press, Cambridge, United Kingdom.

Becker, S.O. and A. Ichino, 2001. Estimation of Average Treatment Effects Based on Propensity Scores. The Stata Journal, 2 (4): 358-377.

Belavadi, N.V. and M.K. Niyogi, 1999. Smallholder Dairy Cooperatives. In: L Falvey and C. Chantalakhana (eds.) Smallholder Dairying in the Tropics. International Livestock Research Institute, Thailand Research Fund, Institute of Land & Food Resources.

Berdegué, J.A., F. Balsevich, L. Flores and T. Reardon, 2005. Central American Supermarkets' Private Standards of Quality and Safety in Procurement of Fresh Fruits and Vegetables. Food Policy, 30 (3): 254-269.

Bernard T., A.S. Taffesse and E.Z. Gabre-Madhin, 2007. Smallholders' Commercialization through Cooperatives: a Diagnostic for Ethiopia. IFPRI Discussion Paper 00722.

Bernard T., A.S. Taffesse and E.Z. Gabre-Madhin, 2008. Impact of Cooperatives on Smallholders' Commercialization Behavior: Evidence from Ethiopia. Agricultural Economics, 39: 1-15.

Berti, P. and W. Leonard, 1998. Demographic and Socio-economic Determinants of Variation in Food and Nutrient Intake in an Andean Community. American Journal of Physical Anthropology, 105: 407-417.

Bertocchi, G. and F. Canova, 2002. Did Colonization Matter for Growth? An Empirical Exploration into the Historical Causes of Africa's Underdevelopment. European Economic Review, 46 (10): 1851-1871.

References

Binswanger, H.P. and T. Nguyen, 2006. Scaling Up Community Driven Development: A Step by Step Approach. the World Bank, Washington DC, USA.

Blackman, A., 2001. Why Don't Lenders Finance High-return Technological Change in Developing-Country Agriculture? American Journal of Agricultural Economics, 83 (4): 1024-1035.

Bonin, J.P., D.C. Jones and L. Putterman, 1993. Theoretical and Emprirical Studies of Producer Cooperatives: Will Ever the Twain Meet? Journal of Economic Literature, XXXI: 1290-1320.

Central Statistical Authority of the Federal Democratic Republic of Ethiopia, 2000-2004. Consumer Expenditure Survey. Statistical Bulletin 200.

Chang, C.C., 2005. The Role of Retail Sector in Agro-food System, Chinese Taipei. Presentation at the Pacific Economic Cooperation Council's Pacific Food System Outlook 2005-6 Annual Meeting in Kun Ming, China, May 11-13.

Chirwa E., A. Dorward, R. Kachule, I. Kumwenda, J. Kydd, N. Poole, C. Poulton, and M. Stockbridge, 2005. Walking Tightropes: Supporting Farmer Organizations for Market Access. Natural Resource Perspectives 99, London: Overseas Development Institute (ODI).

Collion, M.H. and P. Rondot, 1998. Background, Discussions, and Recommendations: Agricultural Producer Organizations, their Contribution to Rural Capacity Building and Poverty Reduction. The World Bank, Washington DC, USA.

Cook, M.L., 1995. The Future of U.S. Agricultural Cooperatives: A Neo-Institutional Approach. American Journal of Agricultural Economics, 77 (5): 1153-1159.

Cook, M.L. and F.R. Chaddad, 2000. Agroindustrialization of the Global Agri-food Economy: Bridging Development Economics and Agri-business Research. Agricultural Economics, 23 (3): 207-218.

Cook, M.L. and M. Chambers, 2007. Role of Agricultural Cooperatives in Global Netchains. Working Paper for the Montpellier Workshop organised by INRA-MOISA and Wageningen University, Wageningen, the Netherlands.

Dadi, L., A. Negassa and S. Franzel, 1992. Marketing Maize and Teff in Western Ethiopia: Implications for Policies Following Market Liberalization. Food Policy, 17 (3): 201-213.

Damiani, O., 2000. The State and Non-traditional Agricultural Exports in Latin America: Results and Lessons of Three Case Studies. Report for the Inter-American Development Bank. Washington, D.C., USA.

Delgado, C., M. Rosegrant, H. Steinfeld, S. Ehui and C. Courbois, 1999. Livestock to 2020, the Next Food Revolution. IFPRI, FAO, ILRI, Discussion paper 28.

Dercon, S., 1995. On Market Integration and Liberalization: Method and Application to Ethiopia. Journal of Development Studies, 32 (1): 112-143.

Dessalegn, G., T.S. Jayne and J.D. Shaffer, 1998. Market Structure, Conduct, and Performance: Constraints of Performance of Ethiopian Grain Markets. Working Paper 9, Grain Market Research Project, Ministry of Economic Development and Cooperation.

D'Haese, M. and G. Van Huylenbroeck, 2005. The Rise of Supermarkets and Changing Expenditure Patterns of Poor Rural households Case Study in the Transkei Area, South Africa. Food Policy, 30 (1): 97-113.

D'Haese, M., G.N. Francesconi and R. Ruben, 2007. Network management for Dairy Productivity and Quality in Ethiopia. In: L. Theuvsen, A. Spiller, M. Peupert and G. Jahn (eds.) Wageningen Academic Publishers, Wageningen, the Netherlands, 185-198.

Cooperation for competition

Di Vico, D., 2008. La Comunitá e il Mercato. Corriere della Sera, Italy, April 15, pp.1.

Dries, L., T. Reardon, and J.F.M. Swinnen, 2004. The Rapid Rise of Supermarkets in Central and Eastern Europe: Implications for the Agrifood Sector and Rural Development. Development Policy Review, 22 (5): 525-556.

Dulfer, E., 1974. Operational Efficiency of Agricultural Cooperatives in Developing Countries. Rome: FAO Agricultural Development Paper No. 96.

Eagleton, D., 2006. Power Hungry: Six Reasons to Regulate Global Food Corporations. Official Report by Actionaid International: www.actionaid.org

Ehret, C., 1979. On the Antiquity of Agriculture in Ethiopia. Journal of African History, 20: 161-177.

Ellis, F., 1988. Peasant Economics, Cambridge University Press, Cambridge, United Kingdom.

Encaoua, D. and A. Jacquemin, 1980. Degree of Monopoly, Indices of concentration and Threat of Entry. International Economic Review, 21 (1): 87-105.

Euromonitor International, 2004. The Global Market for Dairy Products. Stagnito Publishing.

Fafchamps, M., 2005. Market Institutions in Sub-Saharan Africa, Cambridge, Mass.: MIT Press, USA.

Fama, E.F., 1980. Agency Problems and the Theory of the Firm, The Journal of Political Economy, Vol.88: 288-307.

Farina, E., 2002. Consolidation, Multinationalization, and Competition in Brazil: Impacts on Horticulture and Dairy Product Systems. Development Policy Review, 20 (4): 441-45.

FDRE (Federal Democratic Republic of Ethiopia), 1994. Proclamation no. 85/1994 Agricultural Cooperative Societies. Federal Negarit Gazeta, Addis Ababa, Ethiopia.

FDRE (Federal Democratic Republic of Ethiopia), 1998. Proclamation no. 147/1998 to Provide for the Establishment of Cooperative Societies. Federal Negarit Gazeta, Addis Ababa, Ethiopia.

FDRE (Federal Democratic Republic of Ethiopia), 2002. Ethiopia: Sustainable Development and Poverty Reduction Program, Addis Ababa, Ethiopia.

FDRE (Federal Democratic Republic of Ethiopia), 2005. Plan for Accelerated and Sustained Development to End Poverty. Addis Ababa, Ethiopia.

Fitzhugh, H., 1999. Foreword. In: L. Falvey and C. Chantalakhana (eds.) Smallholder Dairying in the Tropics, International Livestock Research Institute, Thailand Research Fund, Institute of Land & Food Resources.

Fuller, F., J.H. Huang and S. Rozelle, 2006. Got milk? The Rapid Rise of China's Dairy Sector and its Future Prospects. Food Policy, 31 (3): 201-215.

Gabre-Madhin, E.Z., 2001. Market Institutions, Transaction Costs, and Social Capital in the Ethiopian Grain Market. Research Report 124, International Food Policy Research Institute: Washington, DC, USA.

Gabre-Madhin E.Z. and I. Goggin, 2005. Does Ethiopia Need a Commodity Exchange? An Integrated Approach to Market Development. EDRI-ESSP Policy Working Paper no.4.

Gebreselassie, S., 2006. Land, Land Policy and Smallholder Agriculture in Ethiopia: Options and Scenarios, Future Agricultures. Ethiopian Economic Association, www.eeaecon.org, Addis Ababa.

Godtland, E.M., E. Sadoulet, A. de Janvry, R. Murgai and O. Ortiz, 2004. The impact of Farmer-Field-Schools on Knowledge and Productivity: A study of Potato Farmers in the Peruvian Andes. Economic Development and Cultural Change, 53: 63-92.

References

Goldman, A. and H. Hino, 2005. Supermarkets vs. Traditional Retail Stores: Diagnosing the Barriers to Supermarkets, Market Share Growth in an Ethnic Minority Community. Journal of Retailing and Consumer Services, 12 (4): 273-284.

Granovetter, M., 1985. Economic Action and Social Structure: The Problem of Embeddedness. American Journal of Sociology, 91: 481-510.

Gutman, G., 2002. Impacts of the Rapid Rise of Supermarkets on Dairy Products Systems in Argentina. Development Policy Review, 20 (4): 409-27.

Hayami, Y. and K. Otsuka, 1992. Beyond the green revolution: agricultural development strategy into new century. In: J.R. Anderson (ed.) Agricultural technology: policy issues for the international community, pp. 35. The World Bank, Washington, DC, USA.

Heckman, J.J., 1979. Sample Selection Bias as a Specification Error, Econometrica, 47: 153-161.

Heckman, J.J., 1998. Matching as an Econometric Evaluation Estimator. Review of Economic Studies, 65: 261-294.

Helmberger, P. and S. Hoos, 1995. Cooperative Enterprise and Organisational Theory. Journal of Cooperatives, 10: 72-86, Reprinted from the Journal of Farm Economics, 44 (May 1963): 275-290.

Hernández, R., T. Reardon, J.A. Berdegué, F. Balsevich and P. Jano, 2004. Acceso de pequeños productores de tomate a los supermercados en Guatemala. Report to USAID, DFID, and the Common Fund for Commodities, under the project RAISE/SPS, Regoverning Markets, and PFID.

Holloway, G., C. Nicholson, C. Delgado, S. Staal and S. Ehui, 2000. Agroindustrialization through Institutional Innovation: Transaction costs, Cooperatives and Milk-Market Development in the East-African Highlands. Agricultural Economics, 23: 279-288.

Hu, D., T. Reardon, S. Rozelle, P. Timmer and H. Wang, 2004. The Emergence of Supermarkets with Chinese Characteristics: Challenges and Opportunities for China's Agricultural Development. Development Policy Review, 22 (4): 557-86.

Humphrey, J., 2007. The Supermarket Revolution in Developing Countries: Tidal Wave or Tough Competitive Struggle? Journal of Economic Geography, 7: 433-450.

Jabbar, M.A., S.K. Ehui and S.J. Staal, 2000. Handbook of Livestock Statistics for Developing Countries. International Livestock Research Institute, Working Paper N.26.

Jansen, H.G.P., 1992. Dairy Consumption in Northern Nigeria. Implications for Development Policies. Food Policy, 17 (3): 114-226.

Jensen, M.C. and W.H. Meckling, 1976. Theory of the Firm: Managerial Behaviour, Agency Costs and Ownership Structure. Journal of Financial Economics, 3 (4): 305-360.

Kaplinsky, R. and M. Morris, 2001. A Handbook for Value Chain Research. http: //sds.ukzn.ac.za/files/handbook_valuechainresearch.pdf.

Karantininis, K. and A. Zago, 2001. Cooperatives and Membership Commitment: Endogenous Membership in Mixed Duopsonies. American Journal of Agricultural Economics, 83 (5): 1266-1272.

Keyzer, M.A. and L. van Wesenbeeck, 2007. Food Aid and Governance. In: E. Bulte and R. Ruben (eds.) Development Economics between Markets and Institutions. Mansholt Publication Series, Vol. 4, Wageningen Academic Publishers, Wageningen, the Netherlands, pp. 183-208.

Kotler, J.P., 1995. Leading Change: Why Transformation Efforts Fail. Harvard Business Review: 59-67.

Lee, J-H. and T. Reardon, 2005. Forward Integration of an Agricultural Cooperative into the Supermarket Sector: The Case of Hanaro Club in Korea. Joint Working Paper, Department of Industrial Economics, Chung-Ang University, Seoul, Korea, and Department of Agricultural Economics, Michigan State University, East Lansing, Michigan, USA.

Lirenso, A., 1993. Grain Marketing Reform in Ethiopia. Ph.D. Dissertation, University of East Anglia, Norwich, United Kingdom.

Luning, P.A., W.J. Marcelis and W.M.F. Jongen, 2006. A Techno-Managerial Approach in Food Quality Management Research. Trends in Food Science & Technology, 17: 378-385.

Mainville, D.Y., D. Zylbersztaijn, E. Farina and T. Reardon, 2005. Determinants of Retailers' Decisions to Use Public or Private Grades and Standards: Evidence from the Fresh Produce Market of São Paulo, Brazil. Food Policy, 30 (3) 334-353.

Mdoe, N. and S. Wiggings, 1996. Dairy Products Demand and Marketing in Kilimanjaro Region, Tanzania. Food Policy, 21 (3): 319-336.

Munckner, H.H., 1988. Principios Cooperativos y Derecho Cooperativa. Friedrich Eberhart Stiftung, Bonn, Germany.

Negassa, A. and T.S. Jayne, 1997. The Response of Ethiopian Grain Markets to Liberalization. Working Paper No. 6. Grain Marketing Research Project, Addis Ababa, Ethiopia.

Neven, D., T. Reardon and R. Hopkins, 2005. Case Studies of Farmer Linking to Dynamic Markets in Southern Africa: The Fort Hare Farmers Group. Michigan State University, East Lansing, MI, USA.

Neven, D., T. Reardon, J. Chege and H. Wang, 2006. Supermarkets and Consumers in Africa: the Case of Nairobi, Kenya. Journal of International Food and Agribusiness Marketing, 18 (3): 103-123.

Nicholson, C.F., 1997. The Impact of Milk Groups in the Shewa and Arsi Regions of Ethiopia: Project Description, Survey Methodology, and Collection Procedures. Mimeograph, Livestock Policy Analysis Project, International Livestock Research Institute, Addis Ababa, Ethiopia.

Nourse, E.G., 1945. The Place of the Cooperative in our National Economy. American Cooperation, , American Institute of Cooperation, Washington DC, USA, pp. 33-39

O'Connor, C.B., 1995. Rural Dairy Technology. ILRI Training Manual n. 1, Addis Ababa, Ethiopia.

Olson, M., 1965. The Logic of Collection Action: Public Goods and the Theory of Groups. Harvard University Press, Cambridge, USA.

Orellana, D. and E. Vasquez, 2004. Guatemala Retail Food Sector Annual, 2004. GAIN Report Number GT4018, USDA Foreign Agricultural Service, Washington, USA.

Pagano, U., 1993. Organizational Equilibria and Institutional Stability. In: S. Bowles, H. Gintis and B. Gustafsson (eds.) Market and Democracy. Cambridge University Press, Cambridge, United Kingdom.

Putterman, L., 1985. Extrinsic Versus Intrinsic Problems in Agricultural Cooperation: Anti-Incentivism in Tanzania and China. Journal of Development Studies 21 (2): 175-204.

Putterman, L. and M. Di Giorgio, 1985. Choice and Efficiency in a Model of Democratic Semi-Collective Agriculture. Oxford Economic Papers, 37 (1): 1-21.

PKF Consulting Ltd, International Research Network, 2005. Dairy Industry in Kenya 2005. www.epzakenya.com.

Rangkuti, F., 2003. Indonesia Food Retail Sector Report 2003. USDA GAIN Report: ID 3028.

References

Ravallion M., 2001. The Mistery of Vanishing Benefits: An Introduction to Impact Evaluation. The World Bank Economic Review, 15 (1): 115-140.

Reardon, T. and C.B. Barrett, 2000. Agroindustrialization, Globalization, and International Development: an Overview of issues, patterns and determinants. Agricultural Economics, 23: 195-205.

Reardon, T. and J.A. Berdegue, 2002. The rapid Rise of Supermarkets in Latin America: Challenges and Opportunities for Development. Development Policy Review, 20 (4): 371-388.

Reardon, T., F. Echanove, R. Cook, N. Tucker and J.A. Berdegué, 2005. The Rise of Supermarkets and the Evolution of their Procurement Systems in Mexico: Focus on Horticulture Products. Working Paper, Michigan State University, East Lansing, USA.

Reardon, T., 2005. Retail Companies as Integrators of Value Chains in Developing Countries: Diffusion, Procurement System Change, and Trade and Development Effects. Commissioned by the German Federal Ministry for Economic Cooperation and Development, Final Report Prepared for Deutsche Gesellschaft fr Technische Zusammenarbeit (GTZ), Germany.

Reardon, T., J. Berdegué and P. Timmer, 2005. Supermarketization of the Emerging Markets of the Pacific Rim: Development and Trade Implications. Journal of Food Distribution Research, 36 (1): 3-12.

Reardon, T., P. Pingali and K. Stamoulis, 2006. Impacts of Agrifood Market Transformation during Globalization on the Poor's Rural Nonfarm Employment: Lessons for Rural Business Development Programs1. Plenary Paper Presented at the 2006 meetings of the International Association of Agricultural Economists, in Queensland, Australia.

Reardon, T., S. Henson and J. Berdegué, 2007. 'Proactive Fast-Tracking' Diffusion of Supermarkets in Developing Countries: Implications for Market Institutions and Trade. Journal of Econometric geography, 7: 1-33.

Rorty, R., 1991. Objectivity, Relativism and Truth. Cambridge University Press, Cambridge, United Kingdom.

Roux, C., P. Le Couedic, S. Durand-Gasselin and F.-M. Luquet, 2000. Consumption Patterns and Food Attitudes of a Sample of 657 Low-Income People in France. Food Policy, 25 (1): 91-103.

Saenz-Segura, F., 2006. Markets and Contracts for Smallholder Pepper Producers: Implications for Production Systems and Resource Management. Chapter Three In: Contract Farming in Costa Rica: Opportunities for Smallholders? Ph.D. Dissertation, Wageningen University, Wageningen, the Netherlands.

Schelhaas, H., 1999. The Dairy Industry in a Changing World. In: L. Falvey and C. Chantalakhana (eds.) Smallholder Dairying in the Tropics. International Livestock Research Institute, Thailand Research Fund, Institute of Land & Food Resources.

Sexton, R.J., 1986. The Formation of Cooperatives: A Game Theoretic Approach with Implications for Cooperative Finance, Decision Making, and Stability. American Journal of Agricultural Economics, 68 (2): 214-25.

Sexton, R. and J. Iskow, 1988. Factors Critical to the Success or Failure of Emerging Agricultural Cooperatives. Giannini Foundation, Information Series No. 88-3, University of California, USA.

Sharma, V.P. and A. Gulati, 2003. Trade Liberalization, Market Reforms and Competitiveness of India Dairy Sector. Markets, Trade and Institutions Division Discussion Paper 61. Washington, DC: International Food Policy Research Institute.

Smith, J. and P. Todd, 2000. Does Matching Overcome Lalonde's Critique of Non-experimental Estimators? Working Paper, Economics Department, University of Pennsylvania, USA.

Spielman, D.J., M.J. Cohen and T.Mogues, 2008. Mobilizing Rural Institutions for Sustainable Livelihoods and Equitable Development: A Case Study of Local Governance and Smallholder Cooperatives in Ethiopia. IFPRI Working Paper, Washington DC, USA.

Staal, S.J., 1995. Peri-urban Dairying and Public Policy in Ethiopia and Kenya: a Comparative Institutional and Economic Analysis. Ph.D. Dissertation, University of Florida, Gainesville, FL, USA.

Staal, S.J., M. Owango, H. Muruiki, M. Kenyanjui, B. Lukuyu, L. Njoroge, D. Njubi, I. Baltenweck, F. Musembi, O. Bwana, K. Muruiki, G. Gichungu, A. Omore and W. Thorpe, 2001. Dairy Systems Characterization of Greater Nairobi Milk Shed. Research Report, SDP (Smallholder Dairy Project), Ministry of Agriculture, Kenya Agricultural Research Institute and International Livestock Research Institute, Nairobi, Kenya.

Staal, S.J., A. Nin Pratt and M. Jabbar, 2006a. Dairy Development for the Resource Poor, Part 2: Kenya and Ethiopia Dairy Development, Case Studies. Pro-Poor Livestock Policy Initiative, Working Paper No. 44-2. Food and Agriculture Organisation, Rome, Italy.

Staal, S.J., 2006b. The role and future of informal and traditional dairy markets in Developing Countries. IGGM&D Dairy Symposium, Rome, Italy.

Staatz, J.M., 1987. Farmers Incentives to take Collective Action via Cooperatives: Transaction Costs Approach. In: J. Royer (ed.) Cooperative Theory: New Approaches. USDA-ACS Service Report No. 18, pp. 87-107.

Strasberg, P.J., T.S. Jayne, T. Yamano, J. Nyoro, D. Karanya and J. Strauss, 1999. Effects of Agricultural Commercialisation on Food Crop Input Use and Productivity in Kenya. MSU International Department of Agricultural Economics Development, Working Paper no. 71.

Sykuta, M.E. and M.L. Cook, 2001. Cooperative and Membership Commitment: A New Institutional Economics Approach to Contracts and Cooperatives. American Journal of Agricultural Economics, 83: 1273-1279.

Taffesse, A.S., B. Fekadu and K. Wamisho, 2006. A Profile of the Ethiopian Economy. ESSP/IFPRI, mimeo.

Taneja, V.K., 1999. Dairy Breeds and Selection. In: L. Falvey and C. Chantalakhana (eds.) Smallholder Dairying in the Tropics. International Livestock Research Institute, Thailand Research Fund, Institute of Land & Food Resources.

Tegegne, A., 2003. Urban and Peri-Urban Dairy Development in Ethiopia: Lessons from the Ada-Liben Woreda Dairy and Dairy Products Marketing Association. Proceedings of Regional Symposium organised by the Amhara Regional Agricultural Research Institute (ARARI), Bahir Dar, Ethiopia.

Tendler, J., 1983. What to Think About Cooperatives: A guide from Bolivia. The Inter-American Foundation, Rosslyn, VA, USA.

Teratanavat, R., V. Salin and N.H. Hooker, 2005. Recall Event Timing: Measures of Managerial Performance in U.S. Meat and Poultry Plants. Agribusiness, 23 (3): 351-373.

References

Thailand Development Research Institute, 2002. The Retail Business in Thailand: Impact of the Large Scale Multinational Corporation Retailers. Working paper, TDRI, Bangkok, Thailand.

The Economist, 2005. Africa's Wall Mart Heads East. January 2005.

The Economist, 2006. Coming to Market. April 2006.

Trail, W.B., 2006. The Rapid Rise of Supermarkets? Development Policy Review, 24 (2): 163-174.

Uphoff, N., 1993. Grassroots Organizations and NGOs in Rural Development: Opportunities with Diminishing States and Expanding Markets. World Development, 21 (4): 607-622.

Veeck, A. and G. Veeck, 2000. Consumer Segmentation and Changing Food Purchase Patterns in Nanjing, PRC. World Development, 28 (3): 457.

Vitaliano, P., 1983. Cooperative Enterprise: an Alternative Conceptual Basis for Analyzing a Complex Institution. American Journal of Agricultural Economics, 65 (5): 1078-1083.

Von Braun, J., 1995. Agricultural Commercialization: Impact on Income and Nutrition and Implications for Policy. Food Policy, 20 (3): 187-202.

Walstra, P., J.T.M. Wouters and T.J. Geurts, 2006. Dairy Science and Technology. Taylor and Francis Group Editors, Boca Raton, USA.

Weatherspoon, D.D. and T. Reardon, 2003. The rise of Supermarkets in Africa: Implications for Agrifood Systems and the Rural Poor. Development Policy Review, 21 (3): 333-355.

Weaver, R.D. and T. Kim, 2001. Contracting for Quality in Supply Chains, Proceedings of EAAE 78th Seminar, Copenhagen, Denmark.

World Bank, 2003. Reaching the Rural Poor, a Renewed Strategy for Rural Development. Discussion Paper, Washington D.C., USA.

World Bank, 2007. Agriculture for Development. World Development Report 2008, Washington D.C., USA.

Yigezu, Z., 2000. DDE's Experience in milk collection, processing and marketing. In The Role of Village Dairy Co-operatives in Dairy Development. Smallholder Dairy Development Project (SDDP) Proceeding, Ministry of Agriculture (MOA). Addis Ababa, Ethiopia.

Zuniga-Arias, G.E., 2007. Quality management and strategic alliances in the mango supply chain from Costa Rica, Wageningen Academic Publishers, Wageningen, the Netherlands, 140 pp.

Summary

Throughout history, rural smallholders have formed various forms of associations (or cooperatives) to confront access-barriers to the market. In the 1960s, many developing countries initiated cooperative development programs, often to facilitate the distribution of subsidised credit and inputs. However, as cooperatives were largely government controlled and staffed, they were often considered as an extended arm of the public sector, rather than institutions or firms owned by the farmers. This form of cooperatives was rarely successful. Still, it is estimated that 250 million farmers in developing countries participate in agricultural cooperatives. Many donors and governments consider agricultural cooperatives to be a fundamental pillar of their rural development policy, as well as a core institution in the process of governance decentralisation and agri-business development. As demonstrated by the growing number of rural cooperatives in the developing world, community empowerment and collective entrepreneurship appears to be particularly viable where infrastructures and markets are poorly developed.

Agriculture is the backbone of the Ethiopian economy, contributing to 48 percent of the gross domestic product. In Ethiopia, 85 percent of the national population (75 million) lives in rural areas under subsistence or semi-subsistence regimes and favourable agro-ecological conditions. In Ethiopia, agricultural cooperatives are a pillar of the national strategy named Agricultural Development-Led Industrialization (ADLI). Agricultural cooperatives, which are seen as key institutions to link national smallholder farmers to emerging supply chains and commodity exchange networks, are advocated by the Ethiopian government to promote the much needed agricultural growth.

To revitalise agricultural growth, the Ethiopian government and various international donors approved, in 2006, the proposal of the International Food Policy Research Institute (IFPRI) to establish and launch the first Ethiopian Commodity Exchange (ECX) by 2008. At the same time, due to rapid urbanisation, and market liberalisation reforms, Ethiopia is witnessing the rapid evolution of the industrial and retail sectors, leading to increasing market integration into supply chains for fresh and perishable food products, and especially for milk.

The scope of this study is to improve the understanding of the role played by cooperative organisations in linking Ethiopian smallholder farmers to emerging supply chains and exchange networks. To do so, this study addresses three major research questions: (a) What are the trends and challenges in the Ethiopian agri-food market? (b) What is the impact of collective action for the competitiveness of Ethiopian farmers? (c) How to promote the competitiveness of Ethiopian rural cooperatives? This study aims at raising awareness of the potential of multidisciplinary research, combining agri-business and development approaches, to analyse the opportunities and challenges of cooperative business in developing countries, and at guiding public-private partnerships towards a pro-poor agro-industrialisation process.

Research objectives are pursued in five analytical chapters based on five different datasets. Quantitative data that form the basis for this study were collected, from the Highland regions of Ethiopia, in the period between 2003 and 2006. The study shows that cooperatives are emerging and growing rapidly in Ethiopia. Here, agricultural cooperatives are mainly considered as associations formed by smallholders to solve both their social and economic problems. To what extent cooperatives meet these expectations depends on their (internal) organisation, and on (external) market and governance characteristics.

The government of Ethiopia is actively promoting the involvement of cooperatives in the newly established commodity exchange (ECX). Using household survey data collected in 2005, Chapter Two analyses the impact of smallholders' cooperatives on agri-commodity (teff, maize, wheat, sesame, and coffee) commercialisation in rural Ethiopia. To do so we examine the factors explaining the degree of commercialisation of cooperative farmers and individual farmers located in major agri-commodity production sites. To eliminate potential diffusion effects between cooperative farmers and farmers that do not belong to cooperatives we select the latter from comparable communities with no cooperatives. Findings suggest that cooperative membership does not have an impact on agri-commodity commercialisation. Only cooperatives that engage in collective marketing activities, such as the collection and sale of members' output, appear to have a significant and positive impact on smallholder commercialisation. However, Chapter Two also shows that Ethiopian agricultural cooperatives, and especially those involved in collective marketing activities, tend to exclude the 'poorest of the poor' (i.e. rural households with little or no land).

Chapter Three evaluates the probability for an Ethiopian agri-cooperative to engage in collective marketing activities over time, given (external) market and governance characteristics. Using a sample of 200 agricultural cooperatives from the Ethiopian Highlands, the analysis suggests that collective marketing activities face cyclical challenges related to increasing competition, and that the average cooperative is an unsustainable form of business organisation. However, empirical results also suggest that in some cases cooperative competitiveness increases over time, especially when a cooperative is established in the regions of Tigray and Amahara, upon the voluntary initiative of farmers.

In Chapter Four the focus shifts towards value chain analysis. In particular, this chapter presents an overview of the trade-offs associated with the evolution of dairy retailers and manufacturing industries in Ethiopia. The analysis makes use of data from 200 consumers in Addis Ababa. Results show that dairy value chains led by modern industries and supermarkets are still scarce in Ethiopia, but are gaining market share as urbanisation and incomes increase. The rise of supermarkets and industries contributes to improve the hygiene of dairy products and outlets, but is associated with increasing concentration of market power into a retail-industrial oligopoly. Oligopoly power translates into high prices and suboptimal nutritional value of final dairy products, and in reduced accessibility of outlets. Consequently, poor and remote consumers tend to be excluded from supermarket outlets and industrial dairy products. As oligopoly power corresponds to oligopsony power, this chapter suggests that smallholder farmers face more constraints, compared to large commercial farmers, in meeting the increasingly stringent specifications of industries and supermarkets.

Using bio-economic data collected in the major Ethiopian milk-shed, Chapter Five evaluates the impact of a marketing cooperative of smallholder farmers, on milk production, productivity, quality and safety. To do so this chapter compares the performance of cooperative farmers and individual farmers within the same area. Findings suggest that membership in marketing cooperatives can have a positive impact on milk production and productivity, no effect on milk hygiene and a negative impact on milk nutritional value.

Using a larger bio-economic dataset from the Ethiopian Highlands, Chapter Six presents practical recommendations to optimise milk quality and safety in national dairy marketing cooperatives, so that Ethiopian smallholder farmers can better compete in the marketplace. The chapter shows that nutritional value and hygiene are extremely poor in the milk produced by Ethiopian cooperative farmers. Poor milk quality is the result of increasing demand pressure from urban consumers, and short-term strategies for profit maximisation by cooperative farmers. However, findings suggest that for a given market and production technology (i.e. herd phenotype), milk hygiene and nutritional value could be still improved through the optimisation of the structure, services, grades and standards adopted by cooperatives.

Chapter Seven discusses the main findings of the study. In particular, the determinants of farmer competitiveness are analysed within the cooperative framework. The typical smallholder farmer in Ethiopia produces mainly for subsistence using a traditional technology which results in little output with a relatively high-quality. Collective action is shown to enable farmers to improve productivity. When collective action involves collective marketing, farmers become more commercial, further improving productivity to the detriment of quality. Finally, options are revealed for marketing cooperatives to follow so that their members can maximise commercialisation and optimise the balance between quality and productivity.

Implications for policy involve the development of rural infrastructures, and more selective public interventions targeting rural collective entrepreneurship. To promote market-oriented cooperatives, public support should not impede with farmers' initiatives to organise collectively but instead formulate appropriate support in the form of managerial capacity building. Finally, the enforcement of better grades and standards to regulate agricultural trade should result from public-private partnerships. In such partnerships the role of the public sector should be the provision of arbitrage over quality control practices, and the role of the private sector (industries and supermarkets) should be the provision of incentives for quality upgrading.

Curriculum Vitae

Gian Nicola Arturo Francesconi was born January 30th, 1976 in Pietrasanta, Tuscany, Italy. In April 2002, he obtained a Laurea (5-year Degree) in Veterinary Science at the University of Pisa, Italy. Subsequently he took a postgraduate course in Rural Development at the Scuola Superiore S.Anna, Pisa University, Italy. At the latter faculty, in November 2003, he also obtained an MSc degree in Food Quality Management and Control. For the writing of his MSc thesis, on the impact of the formation of farmers' cooperatives on milk quality, he carried out six months field research in a village situated in the central Highlands of Ethiopia. His MSc thesis was carried out under the supervision of Simeon Ehui and Azage Tegegne at the Policy Division of the International Livestock Research Institute (ILRI).

In January 2004, he was enrolled as a PhD researcher at the Development Economics group of Wageningen University, the Netherlands, under the supervision of Arie Kuyvenhoven, Ruerd Ruben and Nico Heerink. His PhD programme involved almost 18 months in Wageningen where he successfully completed the doctoral training programme of the Mansholt Graduate School, including courses on development economics, business management, and econometrics. His PhD programme involved also 30 months in Ethiopia, conducting field research sponsored by the International Food Policy Research Institute (IFPRI), under the Supervision of Eleni Gabre-Madhin. During his field work he had the chance to collaborate also with several NGOs, and in particular with SNV (Netherlands Development Organisation).

Currently he is living in Dakar Senegal, where he is consulting the Committee on Sustainability Assessment (COSA), coordinated by the United Nations Conference on Trade and Development (UNCTAD), Solidaridad (Dutch NGO) and other NGOs.

International chains and networks series

Agri-food chains and networks are swiftly moving toward globally interconnected systems with a large variety of complex relationships. This is changing the way food is brought to the market. Currently, even fresh produce can be shipped from halfway around the world at competitive prices. Unfortunately, accompanying diseases and pollution can spread equally rapidly. This requires constant monitoring and immediate responsiveness. In recent years tracking and tracing has therefore become vital in international agri-food chains and networks. This means that integrated production, logistics, information- and innovation systems are needed. To achieve these integrated global supply chains, strategic and cultural alignment, trust and compliance to national and international regulations have become key issues. In this series the following questions are answered: How should chains and networks be designed to effectively respond to the fast globalization of the business environment? And more specificly, How should firms in fast changing transition economies (such as Eastern European and developing countries) be integrated into international food chains and networks?

About the editor

Onno Omta is chaired professor in Business Administration at Wageningen University and Research Centre, the Netherlands. He received an MSc in Biochemistry and a PhD in innovation management, both from the University of Groningen. He is the Editor-in-Chief of The Journal on Chain and Network Science, and he has published numerous articles in leading scientific journals in the field of chains and networks and innovation. He has worked as a consultant and researcher for a large variety of (multinational) technology-based prospector companies within the agri-food industry (e.g. Unilever, VION, Bonduelle, Campina, Friesland Foods, FloraHolland) and in other industries (e.g. SKF, Airbus, Erickson, Exxon, Hilti and Philips).

Guest editors

Ruerd Ruben is professor in development studies and director of the Centre for International Development Issues (CIDIN) at Radboud University Nijmegen, the Netherlands. He holds a PhD in economics from Free University Amsterdam and has published widely on cooperative development and rural organisations, agro-food supply chains and networks, rural land and labour markets, microfinance and sustainable development, including recent articles in World Development, Food Policy, Agricultural Economics, Journal of Agrarian Change, Journal of Chain and Network Science and Supply Chain Management: International Journal.

Arie Kuyvenhoven is (retired) professor in Development Economics and director of Mansholt Graduate School of Social Sciences at Wageningen University, the Netherlands. He started his academic career at the Netherlands School of Economics in Rotterdam, now Erasmus University. He supervised more than 25 PhD theses that covered issues ranging from contract farming, migration, soil conservation, and village modeling. He has been Board member Centre for World Food Studies (SOW), Free University, Amsterdam; Honorary Professor, Nanjing Agricultural University, China; Member Board of Trustees, and later Vice-chair, International Food Policy Research Institute (IFPRI), Washington, D.C., U.S.A. Visiting Professor University of California, Davis; Visiting Professor Cornell University, Ithaca, NY; and membership of the editorial boards of several international journals.

This publication was realised with core financial support from the International Food Policy Research Institute (IFPRI), and additional contributions from SNV (Netherlands Development Organisation) and Wageningen University. Research activities were carried out at IFPRI's regional office for East-Africa, c/o ILRI Addis Ababa, Ethiopia; and at the Development Economics group of Wageningen University, in the Netherlands.

The cover of this volume was designed by the Italian graphic designer Andrea Grillenzoni.

Printed in the United States
by Baker & Taylor Publisher Services